# The Politics of Adapting to Climate Change

Leigh Glover · Mikael Granberg

# The Politics of Adapting to Climate Change

palgrave
macmillan

Leigh Glover
The Centre for Societal Risk Research
Karlstad University
Karlstad, Sweden

Mikael Granberg
The Centre for Societal Risk Research
Karlstad University
Karlstad, Sweden

The Centre for Natural Hazards and
Disaster Science (CNDS)
Uppsala University
Uppsala, Sweden

ISBN 978-3-030-46204-8 ISBN 978-3-030-46205-5 (eBook)
https://doi.org/10.1007/978-3-030-46205-5



This Palgrave Pivot imprint is published by the registered company Springer Nature Switzerland AG
The registered company address is: Gewerbestrasse 11, 6330 Cham, Switzerland

# Preface

This book is the output of the research programme 'Resilience in Sweden: Governing, Social Networks and Learning' (MSB/2016-6855) financed by the research fund of the Swedish Civil Contingency Agency, the research programme 'Reduced climate risks in future housing and dwelling—learning from previous events and societal planning' (P2/14) funded by Länsförsäkringar Alliance research fund and one of the outputs of a more than decade-long collaboration between Leigh Glover and Mikael Granberg financed by different sources over the years. The collaboration started at the Australasian Centre for the Governance & Management of Urban Transport (GAMUT), the University of Melbourne, Australia. At the core of this collaboration has been the exploration of issues connected to climate change and its impacts on society, highlighted by an awareness of the importance of the political aspects of society's effort to address (or not to address) climate change, and its related impacts, in Australia, Sweden and the wider world. This volume is an outcome of this interest. It is also an effort to make a case for greater awareness, understanding and appreciation of the political character and political implications of climate change adaptation policies and plans and the processes of their formulation and delivery. We would like to thank the Centre for Societal Risk Research at Karlstad University, Sweden, for the support we have received working on this book and especially Professor Lars Nyberg, research leader at the Centre, for his comments on an early draft of the manuscript. We also appreciate the input from an anonymous reviewer of the book proposal and the final manuscript for his/her insightful and constructive comments and questions. We also thank the good people at Palgrave Macmillan for their work in developing our ideas and arguments and in the finalizing of this project. Responsibility for the ideas and positions put forth in this book rests, however, firmly with the authors.

Flinders, Australia
Örebro, Sweden
March 2020

Leigh Glover
Mikael Granberg

# Contents

## Part II Adaptation as Politics

# Acronyms

| | |
|---|---|
| CRED | Centre for Research on the Epidemiology of Disasters |
| DRM | Disaster Risk Management |
| DRR | Disaster Risk Reduction |
| EU | European Union |
| FEMA | Federal Emergency Management Agency (USA) |
| GDP | Gross Domestic Product |
| GHG | Greenhouse Gas(es) |
| IFRCRCS | International Federation of Red Cross and Red Crescent Societies |
| ILO | International Labour Organization |
| IPCC | Intergovernmental Panel on Climate Change |
| ISO | International Organization for Standardization |
| ITK | Indigenous Traditional Knowledge |
| NGO | Non-Governmental Organization(s) |
| OECD | Organization for Economic Cooperation and Development |
| SES | Socio-Ecological Systems |
| TEK | Traditional Ecological Knowledge |
| UN | United Nations |
| UNCED | United Nations Conference on Environment and Development |
| UNDP | United Nations Development Programme |
| UNDRR | United Nations Office for Disaster Risk Reduction (formerly UN-ISDR, United Nations International Strategy for Disaster Reduction) |
| UNEP | United Nations Environment Programme |
| UNESCO | United Nations Educational, Scientific and Cultural Organization |
| UNFCCC | United Nations Framework Convention on Climate Change |
| WCED | World Commission on Environment and Development |

# List of Tables

# List of Boxes

# PART I

# Adaptation as a Phenomenon

CHAPTER 1

# The Politics of Adapting to Climate Change

**Abstract** This chapter introduces and lays out the central argument of the book. Adaptation and politics are discussed, and we argue for the need to understand climate change adaptation as a political phenomenon. We examine how adaptation has been defined and understood, explaining how this has changed and evolved. Key concepts and major themes in the policy activities and related scholarship are described, as are the relationships with GHG mitigation and climate change impacts assessment. These concepts include resilience, climate risk and vulnerability. The chapter also explains the chapter arrangement, the different contents therein and provides the reader with a rationale and clear guide to the organization of the book.

**Keywords** Adaptation · Governance · Politics · Practice · Scholarship · Theory

> Climate change can be seen as a classic problem in public policy given the levels of uncertainty, complexity, and ambiguity … Hence climate change risks and public and private responses to them are unstructured in the sense that relevant norms, processes, and facts are contested. (Huitema et al. 2016, p. 37)



L. Glover and M. Granberg,
*The Politics of Adapting to Climate Change*,
https://doi.org/10.1007/978-3-030-46205-5_1

Climate change is a global, disrupting and transformational agent with broad implications for society, managed ecosystems and the natural world (IPCC 1990, 2001, 2007, 2014). Biodiversity loss, already high due to human influences, will be further accelerated by climate change. The range of much valuable flora and fauna will be reduced, as will the diversity of species. Threats to human health will increase with anthropogenic climate change through disruptions to food production, extreme heat and degraded air quality in densely populated cities, increases in vector-borne diseases and other factors. Sea level rise, combined with other elements involved in coastal flooding, threatens the lives and wellbeing of millions of people living on the coastal deltas, especially in Bangladesh, China, Egypt and India. Although these and other impacts will accumulatively impose great costs, there are measures to reduce vulnerabilities, build capacities, plan, prepare and support adaptation measures. Societies and communities around the world are now embarking on a project to prepare for the impacts of climate change and to respond after its impacts are realised. At the same time, uncertainty is prevalent as '… we do not know *and cannot know* exactly what changes – including thresholds and feedbacks – are in motion' (Nightingale et al. 2019, p. 1). It follows that the central questions for climate change adaptation include: Who will make the key adaptation decisions? How will these decisions be made? Who will suffer the losses from climate change? Who will reap the benefits? Accordingly, adapting to climate change involves social agents in distributing the resultant costs and benefits. Society is impacted by a changed climate but society is by no means in a static state. In many ways anthropogenic climate change is a direct result of the development of the modern industrialised society (IPCC 2014). In addition, in parallel with a changing climate, contemporary society is developing in ways that influence vulnerabilities, events impacts, capacities to act and the pathways for future action and decisions (Moloney et al. 2018a). This is further complicated by the array of inconsistencies, multiplicity of meanings and ambiguities that surrounds climate change and related policymaking and governance (Bulkeley 2016). The dynamic of climate change, a changing society, and policymaking and governance involves great challenges and entails comprehensive threats to society.

Responding to these comprehensive threats requires coordinated action at all levels and including all spheres of society and involves '… more than just minor adjustments at the policy scale' (Moloney et al. 2018b, p. 5). These actions include both efforts to reduce greenhouse gas (GHG) emissions to keep the rate of future global warming as low as possible (i.e., mitigation), as well as efforts to counteract the climate change impacts already extant, in the short and in the long term, by assessing risks, preparing for and responding to these impacts (i.e., adaptation). For clarity, in this volume 'adaptation' refers to social responses to actual or anticipated climate change impacts and not to adaptation in the natural world.

From one perspective, mitigation and adaptation pose different challenges for policy design and intervention (Schipper 2006; Granberg and Elander 2007; Biesbrok et al. 2009; Pielke 2009). One key difference between climate change mitigation and adaptation is the stronger role of government, especially central government, in mitigation policies (Head 2014). This is a result of the need for authoritative enforcement of standards '... to target and reduce the several major known sources of GHG emissions' (Head 2014, p. 664). Mitigation effectiveness is independent of place, as it does not matter where the emissions originate, or where the mitigation takes place, as it is the global cumulative effect that is in focus. Adaptation, in contrast, is often considered as a place-based activity focusing on climate-related local and regional impacts, such as extreme weather events, heatwaves and droughts, changing rainfall patterns and sea level rise (Agrawal et al. 2009; Farber 2013; Moloney et al. 2018a; Granberg et al. 2019).

From another perspective, this dichotomous view can be problematized as both mitigation and adaptation are collective action problems lacking clearly identifiable responsibilities (Keskitalo et al. 2016). This highlights the need for involving public actors in ways that also involves steering and coordinating businesses, NGOs, organized interests and individual citizens (Bulkeley 2016; Morrison et al. 2017; Moloney et al. 2018a).

Avoiding ongoing global warming exceeding two degrees Celsius, global GHG emissions must be cut by one-half within 12 years and essentially eliminated by 2050; at the current rate of warming, a 1.5°C warming will occur around 2040 (IPCC 2019). Future climate and climate-related impacts will occur on top of the already heavy toll exacted by climate hazards. Over the period 2008–2017, the 3751 natural hazards (84% were weather-related) recorded by the Centre for Research on the Epidemiology of Disasters (collected in the CRED database), affected two billion people (95% weather-related) and costing US$1658 billion in 141 countries (73% of which was weather-related) (IFRCRCS 2018).

Global climate change will increase the frequency and intensity of extreme weather events and alters climatic conditions worldwide that impacts society and managed natural systems (IPCC 2019). Climate change will bring forth larger storms, higher temperatures and longer droughts, greater coastal flooding and other climate-related hazards causing greater health costs, economic losses, social disruptions, greater loss of life and, if unchecked, the collapse of valuable ecosystem services.

Few would argue with the proposition that mitigating GHG emissions and adapting to climate change impacts are two sides of the same coin, as a cause-effect pairing. Mitigation is closely related to adaptation, as successful or unsuccessful mitigation today will influence the future needs, scope and frequency of climate change adaptation measures (Tol 2005). At the same time, extreme weather events forcing adaptation can push the climate change mitigation agenda forward (Demski et al. 2017). Adaptation measures should also contribute, to a reasonable extent, to wider efforts to reduce GHG emissions either directly or indirectly and adaptation measures that contribute to emissions constitute maladaptation (Barnett and O'Neill 2010; Magnan et al. 2016). Yet, mitigation and adaptation have mostly been separated in the United Nations Framework Convention on Climate Change (UNFCCC), national, regional and local policies and practices, in business and community responses, and in science and scholarship. Partly this arose from early debates where adaptation was cast as a facile and palliative alternative to the urgent and serious matter of securing agreements for emissions reduction (Pielke et al. 2007; Bassett and Fogelman 2013). Adaptation came to be rationalized as the cost of failed mitigation efforts.

Pielke et al. (2007), amongst others, argued that such rationalizations obscured the other existing proximate causes of vulnerability to climate-related disasters (i.e., non-climate factors). Under the UNFCCC's Kyoto Protocol, funding from wealthy nations for adaptation was entirely voluntary (UNFCCC 2008). Worse still, any contributions could be offset by GHG mitigation efforts. Another point of contention is assuming that adaptation is essentially local and inappropriate for actions at other levels (Granberg et al. 2019), an outlook that has contributed to the neglect of adaptation at all levels (Pielke et al. 2007).

Tensions remain over the mixes of mitigation/adaptation measures and associated questions over linkages, synergies, trade-offs and optimality and the challenges of differing time scales, governance scales, policy instruments, institutions and financing options, amongst others (see, Kane and

Shogren 2000; Schipper 2006; Swart and Raes 2007; Morrison et al. 2017). Regardless of the trajectory of future GHG emissions, however, adaptation is essential to reduce vulnerabilities to climate-related hazards and should not be traded-off against mitigation.

So, regardless of the extent of future GHG emissions reductions, accumulated atmospheric emissions to date have resulted in ongoing anthropogenic climate change now ensures climate-related hazards and impacts in the present and the future. As climate-related impacts continue to increase—also for reasons unrelated to climate change, such as population growth in vulnerable locations, environmentally-unsustainable development and growing income inequality (IFRCRCS 2018; UNDRR 2019)—climate change adaptation becomes a social and environmental necessity if such hazards and losses to social and natural systems are to be addressed.

Climate change adaptation is therefore closely related to disaster risk reduction (DRR) and disaster risk management (DRM) as they both deal with lowering the risks of natural hazards prior to, and following their occurrence, with vulnerability reduction as a central and common theme (Birkmann and von Teichman 2010; Birkmann 2013). As already stated above, climate change adaptation takes place in a grey zone with unclear responsibilities that have to be negotiated between societal actors. Furthermore, this grey zone is riddled with uncertainty around what measures and investments that '... are needed today to avoid damage tomorrow and who ought to take responsibility for those actions' (Moloney et al. 2018a, p. 5).

Media coverage of climate change impacts has focused on 'natural' disasters, extreme weather and sea level rise, but the range of impacts is potentially very broad (see, e.g., IPCC 2014). Adaptation has also become bound into the sustainable development agenda as promoted by the United Nations Development Programme (UNDP), United Nations Environment Programme (UNEP), through the aegis of the UNFCCC process and through the sustainable developments goals (SDGs) formulated in the United Nations 2030 Agenda for Sustainable Development (in particular through SDG 13) (UN 2015a). Accordingly, the broad remit of adaptation activity now includes ecosystem-based adaptation, community-based adaptation, indigenous and traditional knowledge, and gender-sensitive approaches. Themes within the adaptation/development realm include agriculture and food security, community safety from climate extremes and disasters, cultural values and systems, ecosystem goods and services, energy sources, human health, livelihood protection, infrastructure and water resources.

Furthermore, climate change impacts can be both direct and indirect (Gough and Meadowcraft 2013; OECD 2014). Direct risks include concrete and tangible impacts, while indirect impacts are spill-over effects from direct climate impacts creating cascading social, ecological and economic outcomes. Predicting these impacts and assessing their riskiness and likelihood of occurring, involves many uncertainties; impact scenarios and predictions are usually accompanied by numerous caveats. Climate change impacts may be unexpected, occur more rapidly than predicted or produce more extreme effects, possibilities that prompt immediate, rather than delayed, adaptations (see, e.g., IPCC 2014). Therefore, several factors favour anticipatory, rather than reactive, adaptations. Precautionary and anticipatory adaptation is likely to be more efficacious and less expensive than delayed, rushed or responsive actions. There may also be short-term social, economic and environmental benefits that can be reaped from more immediate adaptive actions (and existing costs that can be curtailed). Such anticipation can, however, be questioned from a path dependency perspective on the grounds that it may 'lock-in' undesirable existing practices (such as resulting from large-scale infrastructure investments) or that uncertainties over climate impacts make such decisions premature, prompting strategies featuring 'no-regret' options, flexibility and reversibility, and use of 'soft' (i.e., policy-related) adaptations (see, e.g., Fussel 2007; Hallegatte 2009).

Accordingly, it is increasingly evident that deliberate adaptation efforts are necessary and a widening awareness of increasing climate-related hazards has initiated adaptation actions across the social realm with the aim to avoid, ameliorate or take advantage of, these impacts (Jagers et al. 2009; Schipper and Burton 2009; IPCC 2014; Wamsler 2014; Nyberg et al. 2019). Adaptive responses include 'climate-proofing' infrastructure and social systems of production (i.e., increasing tolerance to climatic extremes), improving the adaptability of managed natural systems, managing susceptible production systems more flexibly, and improving knowledge, awareness and understanding of climate risks and vulnerabilities. Climate change adaptation is unavoidable, will continue to grow in societal importance and will necessitate the intervention of states (i.e., public actors on national, regional and local levels), civil society and private enterprise. Despite lingering climate change scepticism and denial by some national governments and political figures, efforts at adaptation are underway in many countries around the world and this action will have to increase in frequency and scope as escalating future climate change impacts create greater risks and consequences.

Climate change adaptation encompasses a vast field of practice. There is much activity underway at the national level and this is set to continue to increase, as promoted by several international initiatives (although some of these pertain to cross-cutting agendas). All of the 175 national signatories to the UNFCCC's Paris Agreement will be reporting on their adaptation activities (UNFCCC 2016). Under the Agreement (see, Article 7), nations are to prepare National Adaptation Plans (NAPs) with the goal of enhancing capacities and resilience and reducing vulnerabilities through measures that contribute to sustainable development. Other initiatives include the UN's SDGs discussed above, the *Addis Ababa Action Agenda on Financing for Development* (UN 2015b) and the *Sendai Framework on Disaster Risk Reduction 2015–2030* (UNISDR 2015) (and the former *Hyogo Framework for Action 2005–2015* [UNISDR 2005]), all of which directly or indirectly promote climate change adaptation.

One indication of the scale of activity is that the UNDP (2016) alone reported that its climate change adaptation portfolio had over 250 projects in over 110 countries and had granted finance of US$1.02 billion. These projects included resilient livelihoods, agriculture, food security, water resources and coastal management, risk management, early warning systems, community-based adaptation, ecosystem-based adaptation, resilient infrastructure and energy. From these programs, over 26 million people benefitted from climate-resilient and risk-management programs, 21 million had improved access to climate information and early warning systems and 1650 km of coastline had been protected as buffer zones for communities at risk from natural hazards (UNDP 2016).

Climate change adaptation activities can include a multitude of actors from different sectors and spheres of society. On the one hand, it has become increasingly clear that local agents—households, communities, non-governmental organizations (NGOs), businesses and local governments—have a critical role (Agrawal et al. 2009; Moloney et al. 2018b). On the other hand, the scope and complexity of the challenge calls for coordinated action and national policy frameworks to deal with large-scale responses and to coordinate, manage and support regional and local action and initiatives (cf., Farber 2013; Sovacool and Brown 2017; Granberg et al. 2019). As Smit and Wandel (2006, p. 289) state:

> While exposures, sensitivities and adaptive capacities are evident at community or local levels, they reflect broader forces, drivers or determinants that shape or influence local level vulnerabilities.

Consequently, growing activity in scale, scope, range of interests and level of resourcing, has been observed among national and local governments, corporations and international agencies (IPCC 2014).

At the same time, climate change adaptation responses and measures vary widely in their social and environmental impacts. Thus, adaptation responses are not equal and individuals, social groups and countries are impacted differently by climate change and, accordingly, need to be treated differently (IPCC 2014). This means that adaptation policy, action and research has to acknowledge the importance of '… how historical dynamics have shaped the current social context through past interventions, disturbances and responses' (Baja and Granberg 2018, p. 126) and how this translates into both benefits and vulnerabilities (cf., Sovacool and Linnér 2016, ch. 7). So, vulnerabilities differences are based in historical development—but contemporary and forward-looking adaptation measures can also produce widely differing societal costs and benefits between individuals, different groups and locations by reinforcing structural resource and power asymmetries in society (cf., the reflections on urban planning in Wilkinson 2012). Ecologically, adaptation can have widely differing effects on resource use, waste production and on ecosystems, habitats and flora and fauna (Pelling 2011; Sovacool and Linnér 2016). Climate change adaptation can, in this sense, be understood as a 'double edged sword' that can maintain and even reinforce path dependency in existing negative social and environmental relations and vulnerabilities, but also has the potential to function as a transformative and reforming force counteracting vulnerabilities to climate change inherent in contemporary societal structures.

Overlaying the politics of adaptation as undertaken, avoided or ignored is the political construct that, put in simple terms: Those responsible for the historical emissions of GHGs are not those who will bear the greatest costs of climate change impacts and who are also likely to have the highest vulnerability to negative climate change impacts. This is one of many complicated ethical and political matters involved in adaptation. Taken together, it takes little imagination to recognize that the outcomes, processes and implementation of adaptation can be considered through a political lens. While acknowledging the critical ethical, philosophical and political debates over the cause of the need for adaptation in the first place, this volume deals with the politics of vulnerability and of adaptation responses.

## Politics in Climate Change Adaptation

In the climate change discourse, adaptation got off to a slow start; in the early international climate agreement negotiations, deliberations, processes and resulting regimes adaptation was largely neglected. Adaptation was essentially absent from the first two rounds of assessment reports produced by the UN scientific advisory body, the Intergovernmental Panel on Climate Change (IPCC) (see, IPCC 1996, 2001). In climate change politics, adaptation has historically attracted the least attention when compared to GHG emissions reduction (mitigation)—not only from policymakers and politicians, but also from social scientists (Ford et al. 2011; Granberg and Glover 2014; Remling and Persson 2014; Taylor 2014; Sovacool et al. 2015). Considering and analysing the political variables of climate change adaptation is, therefore, a relatively new and emerging field of research and scholarship. Typologies of adaptation measures have not typically recognized either the political context, ideology or worldviews embedded in adaptation or the political implications of adaptation measures (e.g., Smit et al. 2000; Biagini et al. 2014).

Our use and understanding of 'politics' in this volume is broad; here, reference to 'politics' includes activities involving power and decision making within a political system and 'the political' refer to the political system and wider social reality (Wiley 2016). In this volume, politics is perceived as a process and practice of governing that takes place through formal and informal institutions shaping political (and other) behaviour (Leftwich 2004; Jagers et al. 2009). Understanding politics and its implications is vital for interpreting the forms and features of societal action. Institutions in this sense are the norms, rules and routines that define, shape and constrain behaviour and action in a specific setting (March and Olsen 1989).

From this perspective, politics is a contest over the framing of ideas (Vink et al. 2013) that impacts the climate change adaptation policy agenda by '… selecting, organising, interpreting and making sense of a complex reality to provide guideposts for knowing, analysing, persuading and acting' (Rein and Schön 1993, p. 146). This means that there are many aspects of adaptation policy that evoke political phenomena, such as the exercising political power, differing values, competing interests, allocating scarce resources, and distributing social and environmental costs.

In many countries, accepting the challenge of climate change adaptation has become an established public policy activity, with governments and official bodies around the world exploring potential impacts, informing communities of risks and making policies and plans for adapting to future climate change. Notwithstanding the fact that adaptation responses to date have largely been coping strategies and plans for increased climate variability and changes in frequency of weather-related disasters, adaptation has become a common feature in development policy and planning systems (i.e., as 'climate risk-management'). Policy initiatives and activities take place at local/regional, national and international levels (Farber 2013; IPCC 2014; Sovacool and Brown 2017; Granberg et al. 2019). These initiatives are sometimes guided by policy from international agencies, such as the UNFCCC, the European Commission, the Organisation for Economic Co operation and Development (OECD), the World Bank, the UNDP, UN-Habitat, USAID and, of course, the IPCC. National measures, according to IPCC (2014, p. 874), reflect:

> … the characteristics of the domestic political structures, socioeconomic conditions, values, and perceptions, as well as development stresses and opportunities.

National activities include new legislation and national policies, financial support to other levels of government, and NAPs. At regional and local levels, the capacity for adaptation is closely connected to the national governance structure that allocate funds, functions and responsibilities between different government levels and between different government actors (IPCC 2014; Moloney et al. 2018a; Granberg et al. 2019). Public initiatives and action sometimes also include partnerships with civic and private sectors. By-and-large, however, governments have approached adaptation as an exercise of centralised planning, delivered through scientific and technocratic approaches (Head 2014; Eriksen et al. 2015). A range of potential political issues for investigation arise from such public sector activities, such as the legitimation of knowledge, the interests of key stakeholders, the expression of power and influence, the role of institutions, transparency and accountability, and the relationship between different societal spheres and government sectors and the extent of subsidiarity.

Who undertakes adaptation, and under what circumstances, are key expressions of the political in adaptation (Bulkeley 2016; Moloney et al. 2018a). Adaptation can be carried out independently, such as by households, firms and organisations, and is necessarily limited by their access to resources and knowledge of available adaptation options. It can also be undertaken collectively and this confers opportunities to distribute both the impacts of climate change and the costs of adaptation between those participating; this also gives access to a greater array of adaptation options. Further, in the decisions over whom undertakes adaptation, abrogation of adaptation responsibilities by higher levels of government effectively transfers the burden of adaptation onto individuals, households, communities and local businesses.

Understanding these political implications requires knowing something of the wider political context in which they occur (Pelling 2011; McEvoy and Fünfgeld 2013) and climate change has come to prominence when the predominant political discourse across developed nations and many developing nations has been neoliberal (Fieldman 2011). There is a dynamic process operating between political factors and adaptation policy and practice, such that adaptation expresses the political and constitutes a political entity with its own identity and political impacts.

This view stands in contrast to much of the earlier thinking on adaptation, which depicted and understood adaptation as a scientific/technical exercise and was taken as being neutral, benign or positive in its political identity and its impacts on society and Nature (Dewulf 2013; Granberg and Glover 2014). As climate change adaptation is given increasing attention in public policy, the research into climate change adaptation policy and concrete responses has expanded from the technological, financial and institutional realms into the wider social realms, including that of politics. Adaptation scholarship has also bridged across to other topics and areas, notably that of development, disaster studies, security and governance. As a result, there are tensions over adaptation policies and practices, such as between technocratic approaches and the engagement of other stakeholders, between those favouring state interventions and those favouring markets, and between social and environmental values and those of markets and economic values.

One concept, already touched upon in this text, that appears throughout the climate change discourse is 'path dependency', notably for explaining the persistence of behaviours, institutions, policy choices, technology choices and other phenomena despite the availability of superior alternatives; i.e., the inability to "shake free" of history (cf., Levin et al. 2012). Drawn largely from economics and now applied widely, conditions favouring path dependency (and taking it beyond the point where diminishing returns to scale might be expected) are large establishment or fixed costs, continual refinements and applied learning, linkage to other systems and the 'self-fulfilling prophecy' effect of anticipated success (see, e.g., David 1985; Arthur 1994; Pierson 2000). Further, as Pierson (2000) points out, many political and institutional decisions are intended to be difficult to change. Oftentimes, reference to conditions being 'locked-in' is an evocation of path dependency.

A new generation of scholarship has taken up these challenges with inquiries into these and related questions (Granberg and Glover 2014; Taylor 2014; Bulkeley 2016; Sovacool and Linnér 2016; Tangney 2017; Klepp and Chavez-Rodriguez 2018; Moloney et al. 2018a), such as: Whose interests are being served? And whose interests are being neglected? What trade-offs have been made? Why have priorities been set in a particular way? What are the social and environmental costs of adaptation? Who contributes to adaptation policy and who is excluded? Is there accountability in decision-making? These studies and critiques have raised issues of political ideology, political values, vested interests and the role of institutions and structures of governance.

An aspect of the adaptation research discourse is whether adaptation pathways should be 'resilient', 'transitional' or socially 'transformational' (Pelling 2011; Moloney et al. 2018a; Granberg et al. 2019). Contemplating this choice necessarily engages the political—for resilience (depending on definition and perspective) can maintain a status quo and uphold existing political circumstances, whilst transformation entails political change. Hence, adaptation is enmeshed in both contemporary politics and those of the future. Therefore, proactive and reactive adaptation approaches present not only technical choices, but also engage political values, interests, institutions and processes.

Directly or indirectly related to the adaptation research discourse is a number of specific concepts (and mainstream concepts that have a particular meaning in the adaptation context) used in both adaptation theory and/or practice (see, Box 1.1). It should be noted, however, that the meaning and understanding of these concepts is often subject to debate and contest—aspects of which are explored in this volume. Readers are therefore advised that the definitions in the box below should be taken as a starting point or background for discussion and consideration. Part of the story of adaptation has been the change and evolution of these concepts and terms—adaptation is a concept that does not stand still, it is very much a moving target (Bassett and Fogelman 2013).

**Box 1.1 Key concepts**

**Adaptive capacity**: That ability of social and natural systems to respond to the impacts of climate change.

**Adaptation limits**: That extent to which adaptation measures can deal with climate change impacts; in some instances, no adaptive measure will be effective, i.e., a hard limit. Soft limits are those where future adaptation measures may be possible but are currently unavailable.

**Climate resilience**: A socio-environmental system's capacity to respond/manage climate change impacts/disruptions (i.e., maintenance of functions); it can apply to loss reduction/avoidance and benefits capture.

**Ecosystem services**: Ecological processes and goods with monetary or non-monetary values, including biodiversity, climate regulation, cultural services (e.g., tourism) and food provision.

**Climate risk**: Usually refers to the likelihood/probability of undesirable climate change impacts occurring and may take into consideration the outcomes of that impact (i.e., hazard plus vulnerability/sensitivity of systems, over time). There are also risks to social and natural systems resulting from adaptation measures.

**Climate vulnerability**: A wide-ranging term usually referring to the extent or degree that a system is susceptible to climate change impacts. Vulnerability is influenced by the system characteristics *and* the impact characteristics in combination.

See, Brooks et al. (2005), IPCC (2014), and Smit and Wandel (2006).

## About This Book

This book addresses the broad issue of the politics of adaptation responses in five steps. As there is a sizable body of scholarship, research and reportage, any researcher can only hope to engage with a relatively small fraction of what is now available. One bibliometric study of climate change publishing in research journals from 1980–2012 identified around 7000 papers on adaptation and another almost 6000 on vulnerability (Haunschild et al. 2016); since the research interest in these themes is increasing; this count is surely much higher now. Publications on adaptation, together with new journals dealing with adaptation themes, are increasing every year. Contemporary digital communication tools enable access to a far greater volume of publications than before such technologies were available, an enabling facility that paradoxically increases the burden for researchers. For the inquiry of this volume—the politics of adaptation—this veritable mountain of research material poses a particular challenge. On the positive side, this abundance of strategies, plans, reports and studies available provides a rich source of material on which this volume has been able to draw. However, the challenge is that this bounty borders on the overwhelming for a thematic study, such as this one. Although a wide-ranging approach has been attempted, no claims are made of having been able to undertake a comprehensive review of the relevant literature.

Divided into two, Part I of this volume addresses adaptation as phenomenon and Part II, as politics. Part I introduces the book, beginning with a description of adaptation and politics, key concepts, purpose of the work, overall argument and the chapter sequence in this chapter. In Chapter 2, the evolution of the adaptation concept is described, highlighting the complexity of the concept, together with the rise of resilience and transformation as a contested political themes and key political issues. Chapter 3 describes the political considerations influencing climate change adaptation politics (as forms of knowledge, socio-economic context and the role of states), together with its links to the disaster and development discourses. In Part II, a case is presented for identifying the political aspects in climate change adaptation and puts forward and supports the argument that understanding climate change adaptation involves understanding its social dimensions, institutional logics and political framing. Chapter 4 provides an account of two key aspects of political context for adaptation, those of governance and sustainability. Chapter 5 identifies the major political themes in adaptation politics, including equity, participation, power, Nature, gender and commons management. Chapter 6 describes and reviews studies that have analysed adaptation politics, grouped thematically as institutional, political economy, environmental justice and political ecology works. This chapter draws on the two contextual themes of Chapter 4 and describes how the major adaptation themes of Chapter 5 have been addressed in the studies of adaptation politics; we show how which themes have been selected and addressed by different political approaches—and how these give rise to these contrasting four political outlooks. In the final chapter (Chapter 7), factors shaping adaptation's evolution are described, along with speculations over its future.

The purpose of this book is to:

1. Contribute theoretically to the body of knowledge around the politics of climate change adaptation
2. Contribute to an understanding of climate change adaptation as a political activity
3. Identify and critically assess the implication of politics on the study and practice of climate change adaptation, and
4. Develop an analytical approach to analysing and understanding climate change adaptation focusing on governance, policy, political values and ideologies.

This book examines the political themes and perspectives related to, and influencing, climate change adaptation. A basic case for examining and addressing the politics of adaptation rests on the goals of understanding and of making a social and environmental contribution. From the educative perspective, this volume seeks to better understand and appreciate the state of knowledge and to place it within a broader scholastic context. It is hoped that such knowledge will provide insights that will better ensure effective adaptation policies and practices. As such, this should serve the interests of normative social and environmental goals in adaptation practices (and help reduce maladaptation due to political factors). In a wider context, it is hoped this will assist the project of progressive politics, covering both its egalitarian and representative goals.

Accordingly, this work intends to provide an informed primer on the politics of adaptation, something neglected in the current scholarship and literature. Although there are now dedicated political works on adaptation, these are usually quite specialized in focus and theme. As a result, basic questions have been largely overlooked: Why are these politics important? What do they mean? What are the implications? This task begins in the next chapter with a description of the understanding of adaptation as concept and as a practice.

## References

Agrawal, A., Kononen, M., & Perrin, N. (2009). *The role of local institutions in adaptation to climate change*. Washington: The World Bank.

Arthur, W. B. (1994). *Increasing returns and path dependence in the economy*. Ann Arbor, MI: University of Michigan Press.

Baja, K., & Granberg, M. (2018). From engagement to empowerment: Climate change and resilience in the city of Baltimore. In S. Moloney, H. Fünfgeld, & M. Granberg (Eds.), *Local action on climate change: Opportunities and constraints* (pp. 126–145). London and New York: Routledge.

Barnett, J., & O'Neill, S. J. (2010). Maladaptation. *Global Environmental Change, 20*(2), 211–213.

Bassett, T. J., & Fogelman, C. (2013). Déjà vu or something new? The adaptation concept in the climate change literature. *Geoforum, 48*, 42–53.

Biagini, B., Bierbaum, R., Stults, M., Dobardzik, S., & McNeeley, S. M. (2014). A typology of adaptation actions: A global look at climate adaptation actions financed through the global environment facility. *Global Environmental Change, 25*, 97–108.

Biesbrok, G. R., Swart, R. J., & van der Knaap, W. G. M. (2009). The mitigation–adaptation dichotomy and the role of spatial planning. *Habitat International, 33*(3), 230–237.

Birkmann, J. (2013). *Measuring vulnerability to natural hazards—Towards disaster resilient societies* (2nd ed.). New York: United Nations University Press.

Birkmann, J., & von Teichman, K. (2010). Integrating disaster risk reduction and climate change adaptation: Key challenges—scales, knowledge, and norms. *Sustainability Science, 5*(2), 171–184.

Brooks, N., Adger, W. N., & Kelly, P. M. (2005). The determinants of vulnerability and adaptive capacity at the national level and the implications for adaptation. *Global Environmental Change, 15*, 151–163.

Bulkeley, H. (2016). *Accomplishing climate governance*. Cambridge: Cambridge University Press.

David, P. A. (1985). Clio and the economics of QWERTY. *American Economic Review, 75*(2), 332–337.

Demski, C., Capstick, S., Pidgeon, N., Gennaro Sposato, R., & Spence, A. (2017). Experience of extreme weather affects climate change mitigation and adaptation responses. *Climatic Change, 140,* 149–164.

Dewulf, A. (2013). Contrasting frames in policy debates on climate change adaptation. *WIREs Climate Change, 4,* 321–330.

Eriksen, S. H., Nightingale, A. J., & Eakin, H. (2015). Reframing adaptation: The political nature of climate change. *Global Environmental Change, 35*(1), 523–533.

Farber, D. A. (2013). Issues of scale in climate governance. In J. S. Dryzek, R. B. Norgaard, & D. Schlosberg (Eds.), *The Oxford handbook of climate change and society* (Paperback ed., pp. 479–503). Oxford: Oxford University Press.

Fieldman, G. (2011). Neoliberalism, the production of vulnerability and the hobbled state: Systemic barriers to climate adaptation. *Climate and Development, 3*(2), 159–174.

Ford, J. D., Berrang-Ford, L., & Paterson, J. (2011). A systematic review of observed climate change adaptation in developed nations. *Climatic Change, 102*(2), 327–336.

Fussel, H.-M. (2007). Adaptation planning for climate change: Concepts, assessment approaches, and key lessons. *Sustainability Science, 2,* 265–275.

Gough, I., & Meadowcraft, J. (2013). Decarbonizing the welfare state. In J. S. Dryzek, R. B. Norgaard, & D. Schlosberg (Eds.), *The Oxford handbook of climate change and society* (Paperback ed., pp. 490–503). Oxford: Oxford University Press.

Granberg, M., Bosomworth, K., Moloney, S., Kristianssen, A.-C., & Fünfgeld, H. (2019). Can regional-scale governance and planning support transformative adaptation? A study of two places. *Sustainability, 11*(24), 6978. https://doi.org/10.3390/su11246978.

Granberg, M., & Elander, I. (2007). Local governance and climate change: Reflections on the Swedish experience. *Local Environment, 12*(5), 537–548.

Granberg, M., & Glover, L. (2014). Adaptation and maladaptation in Australian national climate change policy. *Journal of Environmental Policy & Planning, 16*(2), 147–159.

Hallegatte, S. (2009). Strategies to adapt to an uncertain climate change. *Global Environmental Change, 19*(2), 240–247.

Haunschild, R., Bornmann, L., & Marx, W. (2016). Climate change research in view of bibliometrics. *PLoS ONE, 11*(7), e0160393. https://doi.org/10.1371/journal.pone.0160393.

Head, B. (2014). Evidence, uncertainty, and wicked problems in climate change decision making in Australia. *Environment and Planning C: Government and Policy, 32,* 663–679.

Huitema, D., Adger, W. N., Berkhout, F., Massey, E., Mazmanian, D., Munaretto, S., et al. (2016). The governance of adaptation: Choices, reasons, and effects. *Ecology and Society, 21*(3), 37.

IFRCRCS. (2018). *World disasters report 2018*. Geneva: IFRCRCS.

IPCC. (1990). *Climate change 1990: The IPCC impacts assessment*. Canberra: Australian Government Publishing Service.

IPCC. (1996). *Climate change 1995: Impacts, adaptations, and mitigation of climate change: Scientific-technical analyses*. Cambridge: Cambridge University Press.

IPCC. (2001). *Climate change 2001: Impacts, adaptation and vulnerability*. Cambridge: Cambridge University Press.

IPCC. (2007). *Climate change 2007: Impacts, adaptation and vulnerability*. Cambridge: Cambridge University Press.

IPCC. (2014). *Climate change 2014: Impacts, adaptation, and vulnerability*. Cambridge: Cambridge University Press.

IPCC. (2019). *Global warming of 1.5°C*. Geneva: IPCC.

Jagers, S., Duus-Otterström, G., & Stripple, J. (2009). Fair and feasible climate change adaptation. *Statsvetenskaplig Tidskrift, 111*(1), 53–58.

Kane, S., & Shogren, J. F. (2000). Linking adaptation and mitigation in climate change policy. In S. M. Kane & G. W. Yohe (Eds.), *Societal adaptation to climate variability and change* (pp. 75–102). Dordrecht: Springer.

Keskitalo, E. C. H., Juhola, S., Baron, N., Fyhn, H., & Klein, J. (2016). Implementing local climate change adaptation and mitigation actions: The role of various policy instruments in a multi-level governance context. *Climate, 4*(7). https://doi.org/10.3390/cli4010007.

Klepp, S., & Chavez-Rodriguez, L. (Eds.). (2018). *A critical approach to climate change adaptation*. London: Routledge.

Leftwich, A. (2004). Thinking politically: On the politics of Politics. In A. Leftwich (Ed.), *What is politics? The activity and its study* (pp. 1–22). Cambridge: Polity Press.

Levin, K., Cashore, B., Bernstein, S., & Auld, G. (2012). Overcoming the tragedy of super wicked problems: Constraining our future selves to ameliorate global climate change. *Policy Sciences, 45*, 123–152.

Magnan, A., Schipper, E. L. F., Burkett, M., Bharwani, S., Burton, I., Eriksen, S., et al. (2016). Addressing the risk of maladaptation to climate change. *Wiley Interdisciplinary Reviews: Climate Change, 7*(5), 646–665.

March, J. G., & Olsen, J. P. (1989). *Rediscovering institutions: The organizational basis of politics*. New York: The Free Press.

McEvoy, D., & Fünfgeld, H. (2013). Resilience and climate change adaptation: The importance of framing. *Planning Practice & Research, 28*(3), 280–293.

Moloney, S., Fünfgeld, H., & Granberg, M. (2018a). Climate change responses the global to local scale: An overview. In S. Moloney, H. Fünfgeld, &

M. Granberg (Eds.), *Local action on climate change: Opportunities and constraints* (pp. 1–16). London and New York: Routledge.
Moloney, S., Fünfgeld, H., & Granberg, M. (Eds.). (2018b). *Local action on climate change: Opportunities and constraints*. London and New York: Routledge.
Morrison, T. H., Adger, W. N., Brown, K., Lemos, M.-C., Huitema, D., & Hughes, T. P. (2017). Mitigation and adaptation in polycentric systems: Sources of power in the pursuit of collective goals. *WIRes Climate Change, 8*(September/October). https://doi.org/10.1002/wcc.479.
Nightingale, A. J., Eriksen, S., Taylor, M., Forsythe, T., Pelling, M., Newsham, A., et al. (2019). Beyond technical fixes: Climate solutions and the great derangement. *Climate and Development*. https://doi.org/10.1080/17565529.2019.1624495.
Nyberg, L., Hakkarainen, H., Blumenthal, B., & Moberg, J.-O. (2019). *Konsekvenser av sommarskyfall i Sverige under åren 2009–2018 – analys av rapportering i dagstidningar* (Rapport 2019:2). Karlstad: The Centre for Climate and Safety, Karlstad University.
OECD. (2014). *Seine basin, île-de-France, 2014: Resilience to major floods*. Paris: OECD.
Pelling, M. (2011). *Adaptation to climate change: From resilience to transformation*. London: Routledge.
Pielke, R. A. (2009). Rethinking the role of adaptation in climate policy. In E. L. F. Schipper & I. Burton (Eds.), *The Earthscan reader in adaptation to climate change* (pp. 345–358). London: Earthscan.
Pielke, R. A., Prins, G., Rayner, S., & Sarewitz, D. (2007). Lifting the taboo on adaptation. *Nature, 445*(8), 597–598.
Pierson, P. (2000). Increasing returns, path dependence, and the study of politics. *American Political Science Review, 92*(2), 252–267.
Rein, M., & Schön, D. (1993). Reframing policy discourse. In F. Fischer & J. Forester (Eds.), *The argumentative turn in policy analysis and planning* (pp. 145–166). London: Duke University Press.
Remling, E., & Persson, Å. (2014). Who is adaptation for? Vulnerability and adaptation benefits in proposals approved by the UNFCCC Adaptation Fund. *Climate and Development, 7*(1), 16–34.
Schipper, E. L. F. (2006). Conceptual history of adaptation in the UNFCCC process. *Review of European Community and International Environmental Law, 15*(1), 82–92.
Schipper, E. L. F., & Burton, I. (Eds.). (2009). *The Earthscan reader in adaptation to climate change*. London: Earthscan.
Smit, B., Burton, I., Klein, R. J. T., & Wandel, J. (2000). An anatomy of adaptation to climate change and variability. *Climatic Change, 45*(1), 223–251.

Smit, B., & Wandel, J. (2006). Adaptation, adaptive capacity and vulnerability. *Global Environmental Change, 16*, 282–292.
Sovacool, B. K., & Brown, M. A. (2017). Scaling the policy response to climate change. *Policy and Society, 27*, 317–328.
Sovacool, B. K., & Linnér, B.-O. (2016). *The political economy of climate change adaptation*. Basingstoke: Palgrave Macmillan.
Sovacool, B. K., Linnér, B.-O., & Goosite, M. E. (2015). The political economy of climate change adaptation. *Natural Climate Change, 5*(July), 616–618.
Swart, R., & Raes, F. (2007). Making integration of adaptation and mitigation work: Mainstreaming into sustainable development policies? *Climate Policy, 7*(4), 288–303.
Tangney, P. (2017). *Climate adaptation policy and evidence: Understanding the tensions between politics and expertise in public policy*. London: Routledge.
Taylor, M. (2014). *The political ecology of climate change adaptation: Livelihoods, agrarian change and the conflicts of development*. London: Routledge.
Tol, R. S. J. (2005). Adaptation and mitigation: Trade-offs in substance and methods. *Environmental Science & Policy, 8*, 572–578.
UN. (2015a). *Transforming our world: The 2030 agenda for sustainable development*. New York: UN.
UN. (2015b). *Addis Ababa action agenda of the third international conference on financing for development*. New York: UN.
UNDP. (2016). *Scaling up climate action to achieve sustainable development goals*. New York: UNDP.
UNDRR. (2019). *Global assessment report on disaster risk reduction*. Geneva: UNDRR.
UNFCCC. (2008). *Kyoto protocol to the united nations framework convention on climate change*. New York: United Nations.
UNFCCC. (2016). *Paris agreement*. New York: United Nations.
UNISDR. (2005). *Hyogo framework for action 2005–2015: Building the resilience of nations and communities to disasters*. Geneva: UNISDR.
UNISDR. (2015). *Sendai framework for disaster risk reduction 2015–2030*. Geneva: UNISDR.
Vink, M. J., Dewulf, A., & Termeer, C. (2013). The role of knowledge and power in climate change adaptation governance: A systematic literature review. *Ecology and Society, 18*(4), 46.
Wamsler, C. (2014). *Cities, disaster risk, and adaptation*. London: Routledge.
Wiley, J. (2016). *Politics and the concept of the political: The political imagination*. New York: Routledge.
Wilkinson, C. (2012). Urban resilience: What does it mean in planning practice? *Planning Theory & Practice, 13*(2), 319–324.

CHAPTER 2

# Climate Change Adaptation as an Idea and a Practice

**Abstract** In this chapter, we provide a general description of climate change adaptation as an idea and a set of practices. It begins with an account of the changing perspectives of adaptation and an overview of key aspects, including sectors of most activity, goals, measures and barriers. The relationship to two related discourses, namely disaster risk reduction and sustainable development, is examined and the influence on adaptation is discussed. Three adaptation frameworks are identified and discussed: Resilience, Transition and Transformation. A critical review is undertaken of the resilience framework and the issue of whether or not it constitutes a means to continue the status quo is examined in a political context. These materials provide a foundation for the examination of emerging political influences on adaptation in the following chapter.

**Keywords** Adaptation · Resilience · Transformation · Transition · Disaster risk · Sustainable development



L. Glover and M. Granberg,
*The Politics of Adapting to Climate Change*,
https://doi.org/10.1007/978-3-030-46205-5_2

## Developing Perspectives on Adaptation

Long-term climate variability is a natural phenomenon and since the beginning of time, humans and society have been in a perpetual and permanent process of responding and adapting to climate and its associated systems (Adger et al. 2009; Hulme 2009; Pelling 2011). Contemporary climate change, however, entails events and impacts outside this accumulated experience as it introduces '... new risks and increasing fluctuations in resources across time and space, exacerbating existing vulnerability' (Eriksen et al. 2015, p. 523). Hence, anthropogenic climate change is disrupting established ways of experiencing and understanding climate and adaptation as it '... poses new challenges to individuals, organizations, and societies' (Bauer et al. 2012, p. 280). Potential impacts stretch across the span of social concerns, including the economy, environmental conditions, health and welfare, and infrastructure and productive systems.

Arguably, a changing climate that leads to extreme weather and changes in climatic conditions with substantial impacts on society drives demands for society to adjust and adapt (Schipper and Burton 2009; IPCC 2014). As suggested above, there are features of contemporary climate change that make adaptation far more 'challenging' than historical weather and climate adaptations. Fundamentally, contemporary climate change is unprecedented in the context of modern industrialised societies (Pelling 2011; IPCC 2014). Future climate change and climate-related phenomena (the forecasts of which can take the form of scenarios) are inherently uncertain, differing meaningfully from responses to adapt to the current climate and related systems (Hulme et al. 1999; Rosenzweig et al. 2008). An implication is that the time horizons of climate change limit the usefulness of scientific knowledge as a basis for decision making and increasing the importance of value judgements. Values are shaped by the structures and norms of governance, political and social systems (Paavola and Adger 2006). Increasing importance of values highlights the need to focus on questions of power as it is vital to think about '... who it is that determines the principles upon which adaptive choices are made as much as the nature of the decisions themselves' (Pelling 2011, p. 47).

Anthropogenic climate change overlays natural climatic variations, making long-term trends difficult to predict and attributing specific hazards, events and climate change impacts are contentious (Rosenzweig et al. 2008). Pelling (2011, p. 25) describes the problem as:

> The vastness of climate change and the multitude of pathways through which it can affect life and wellbeing for any individual or organization make it almost impossible for 'climate change' in a wholistic sense to be the target of adaptation.

As such, adaptation measures are rarely stand-alone but rather, supplement existing plans and policies and are therefore usually incremental in their effects. This is a problem, and a challenge, for adaptation.

Adaptation to climate change as a deliberate public policy action is only as old—more or less—as climate change has been an international issue (Moser and Ekstrom 2010). As discussed in Chapter 1, however, adaptation was downplayed through the early years as it was regarded by some as being contrary to climate change mitigation (Schipper 2006; Pielke et al. 2007; Bassett and Fogelman 2013). Over time, however, the rationales for undertaking adaptation have become well-established, especially in light of climate change as effectively a permanent feature of social life on earth (Berrang-Ford et al. 2011; Eriksen et al. 2015). IPCC (2014) argues that precaution is justified because anticipatory actions are less expensive and less disruptive than reactive responses in the face of more acute and dramatic future climate impacts. More compellingly, the IPCC (2014) observes that, climate change-related impacts may occur abruptly and escape prediction, thereby eliminating any rationalisations for delaying adaptation preparations.

As stated in Chapter 1, adaptation, as a concept and a practice, is almost impossibly broad, engaging as it does the economic, social, environmental, technological, legislative and, as argued here, the political (for an overview of adaptation actions, scales and actors involved, see Moloney et al. 2018). Adaptation occurs in a temporal sense in relation to climate and climate-related impacts, it can be preventive, preparatory, reactive and recuperative (i.e., recovery activities). In scope, adaptation ranges from the local to the global. Thus, the societal challenge that adaptive action on climate change offers is composite of short- and long-term dimensions, of different scales of response and of geographical focus. Considering the potential (and already occurring) costs of climate change, any listing of potential impacts is almost only limited by the scope of inquiry undertaken (Fankhauser et al. 1999; Kahn 2016). One implication is that conventional economic cost-benefit approaches for informing public policy are unsuitable. More broadly, estimating the costs of adaptation at global, national, local and other scales has proved problematic

(Parry et al. 2009). Vulnerability of social systems to climate change impacts is potentially enormous—both in the range of impacts and in their potential scale.

As adaptation can be a goal, a process or an outcome, adaptation typologies are diverse. These typologies include infrastructure/technological, institutional, communal and ecological/ecosystem, behavioural, capacity-building, financing, management, planning, warning and observation measures and, although somewhat neglected as an adaptation in its own right, research (Biagini et al. 2014; IPCC 2014). While adaptation is usually associated with a long-term perspective, it is also related to short-term coping strategies. Coping is about handling the more immediate stresses and disturbances caused by external factors, such as climate change (Jabareen 2013). As coping is more focused on short-term adaptive measures it cannot be understood as more continuous and long-term climate change adaptation. At the same time, adaptation is often viewed as being either:

- *Reactive* (i.e., following an event) or
- *Proactive* (i.e., precautionary and preparatory), but it can also be deliberately
- *Inactive* (i.e., based on perceptions that adaptation may be deemed unnecessary, infeasible or economically unattractive).

Table 2.1, gives an overview of key aspects of adaptation, capturing this complexity.

As Table 2.1 indicates, adaptation options are of a great magnitude, differentiated by a wide range of factors, including location, stakeholders, timeframe, impact type, cost of responses, cost of potential losses and likelihood of future impacts. Accordingly, adaptation can be understood in a number of ways and these have changed over time, as the following overview of the IPCC definitions demonstrates.

Created to support the UN climate change negotiations process, the IPCC provides an account of the relevant research as compiled by researchers and government officials (Oreskes 2005). Adaptation got off to a poor start and is not covered in the first assessment report (IPCC 1990). In the second assessment report (a more substantial effort all around), adaptability is defined as (IPCC 1996, p. 28):

**Table 2.1** Overview of climate change adaptation: Key aspects

| | |
|---|---|
| Prominent sectors of adaptation activity | Agriculture<br>Coastal zones<br>Ecosystems/ecosystem services<br>Human health<br>Financial services/insurance<br>Cities<br>Settlements/industries<br>Water resources |
| Goals of adaptation | Enhanced adaptability<br>Enhanced robustness<br>Flexibility increases<br>Preparation improvements<br>Vulnerability reduction |
| Adaptation options/measures | Behaviour change<br>Education<br>Engineering<br>Institutions<br>Law/regulation/administration<br>Market-based measures<br>Planning |
| Barriers to adaptation | Cultural/social<br>Educational<br>Financial/economic resourcing<br>Informational<br>Institutional<br>Technological<br>Politics |

See, Adger et al. (2013), IPCC (2014), Kahn (2016), and Pelling (2011)

> … the degree to which adjustments are possible in practices, processes, or structures of systems to projected or actual changes of climate; adaptation can be spontaneous or planned, and can be carried out in response to or in anticipation of changes in conditions.

Basically, this approach to adaptation is an ecological one; in practice, this matters little as the issue is scarcely covered in the report. By the third IPCC assessment report, the definition has evolved somewhat (IPCC 2001, p. 982):

> Adjustment in natural or human systems in response to actual or expected climatic stimuli or their effects, which moderates harm or exploits beneficial opportunities. Various types of adaptation can be distinguished, including anticipatory and reactive adaptation, private and public adaptation, and autonomous and planned adaptation.

Although social adaptations are deemed identical to ecosystem adaptations, the potential benefits of climate change are recognised. More importantly, the report draws on the growing scholarly literature and could be said, belatedly, to have recognised its significance. No definitional change is made in fourth assessment report (IPCC 2007). In 2012, the IPCC produced a special report on extreme events, disasters and adaptation and defined adaptation as (IPCC 2012, p. 556):

> In human systems, the process of adjustment to actual or expected climate and its effects, in order to moderate harm or exploit beneficial opportunities. In natural systems, the process of adjustment to actual climate and its effects; human intervention may facilitate adjustment to expected climate.

In the fifth assessment report, the adaptation definition shows changes in outlook (IPCC 2014, p. 1758):

> The process of adjustment to actual or expected climate and its effects. In human systems, adaptation seeks to moderate or avoid harm or exploit beneficial opportunities. In some natural systems, human intervention may facilitate adjustment to expected climate and its effects.

It also discusses the incremental adaptation that unfolds within a system with the purpose of maintaining a system but also more radical transformational adaptation that '… changes the fundamental attributes of a system in response to climate and its effects' (IPCC 2014, p. 1758). Furthermore, the report picks up the themes of mainstreaming adaptation and the connections to sustainable development.

A sometime overlooked aspect of beneficial adaptation is that enabling better responses to routine climate impacts and changes institutional practices in ways that create 'co-benefits' for society. Co-benefits can arise in, e.g., public health (Elmquist et al. 2019), local government services (Fünfgeld 2015) and through transformational outcomes (Pelling 2011). Following the precautionary principle (i.e., a result in benefits whether or not anticipated climate change impacts materialize), adaptation activities can be justified on their co-benefits alone. Finally, as explored below, adaptation measures can be socially progressive and environmentally desirable; indeed, they may become an essential component of GHG emissions mitigation efforts (e.g., Nyong et al. 2007; Verchot et al. 2007; Ayers and Huq 2008).

Adaptation is concentrated in those socio-economic sectors characterised by a high degree of management and control, such as agriculture, managed ecosystem services, urban settlements and coastal areas. Conversely, those areas highlighted as being susceptible to great disruption by climate change over which there is little management or control, such as whole ecosystems, marine areas and many ecosystem services, are subject to little adaptation (IPCC 2014). Further, adaptation has, historically, been highly tied to the predictive capacities of climate change science (Tembya et al. 2016; Nightingale et al. 2019). Of concern is that adaptation appears to be neither feasible or likely for the climate change impacts with 'very low probabilities/very high impacts' or the unpredicted 'surprise' events (i.e., major abrupt climatic changes and impacts).

Adaptation responses can engage much of the spectrum of modern social functions (Patterson and Huitema 2019), including behaviour change, education, engineering, government institutions, law/regulation/administration and planning. Typically, four main types of responses are recognized (see, e.g., Pelling 2011; IPCC 2014; Moloney et al. 2018). *Infrastructure* responses focus on infrastructure and technology and are the most common adaptation approach. *Institutional* responses are typically the responses by governments, covering planning, regulations and the roles and services for agencies that can include awareness-raising, education, emergency procedures, warning and advisory services, and research and knowledge creation. *Social* or *community* responses embrace social networks, community-based resource management and community organizations. *Ecosystem* responses refer to managed ecosystems and ecosystem services, and includes species and habitat conservation and use of ecosystem services. For clarity, ecosystem services encompass that enormous span of goods and services underpinning all social life and human survival (Burkhard and Maes 2017) (sometimes categorised into provisioning, regulating, supporting and cultural forms); managed ecosystems provide essential goods and services and other benefits to society through deliberate ecosystem manipulations (often entailing conservation, restoration and sustainability), including agriculture, water harvesting, waste disposal and the like.

Adaptation can also lead to negative outcomes. If adaptation measures fail or have counter-productive effects, they can be said to be 'maladaptations' (essentially applying a concept from evolutionary biology to the socio-environmental realm) (Barnett and O'Neill 2010; Granberg and Glover 2014). Early uses of the maladaptation concept were relatively straightforward and referred to adaptation efforts that increased vulnerability to climate change impacts, usually in anthropological and historical accounts of environmental crises. Magnan et al. (2016) canvassed three complications in defining maladaptation: (1) Whether it must result exclusively from adaptation efforts or from other contributing factors, (2) Whether it must occur in the same location as the adaption measure and (3) Whether it can be reasonably attributed over longer time frames (i.e., unanticipated consequences).

Barnett and O'Neil (2010) considered a broader approach to assessing adaptation measures by including reference to other systems, other sectors and other social groups and identified five types of maladaptation: (1) Increases GHG emissions, (2) Socially regressive by disproportionally burdening the most vulnerable, (3) High opportunity costs, (4) Reduced incentive to adapt and (5) Contributing to path dependency so that future choices are reduced. Subjectivity comes into play here; suitable adaptations to one group may be maladaptations to another. As a concept, maladaptation remains elusive as there is seldom any analysis of the experiences among different actors and over geographical and temporal scales (Neset et al. 2019). Hence, there are considerable differences in opinion as to how far and wide, and into the future, adaptation effects need to be assessed to judge whether maladaptation has taken place or not (Juhola et al. 2016).

The relationship between adaptation and knowledge is central (Boezeman 2015; Strauss 2015; Tembya et al. 2016; Palutikof et al. 2019). Adaptation responses draw on a knowledge base of local, traditional, indigenous and scientific information that must be collected, organised, processed, analysed, distributed and utilised. A part of this activity is the need to identify knowledge requirements. This knowledge includes that required to assess vulnerabilities, risks and existing capacities for adaptation. Policy formulation and planning processes involves many steps, including stakeholder engagement and participation, inter-agency coordination, financing and other considerations. Usually options are developed and evaluated as part of these processes. Specialist institutions and bodies may need to be established to manage programs and projects. Implementation also involves a range of practical issues, including stakeholder engagement, project delivery, information exchanges, promotion and marketing. Monitoring, reviewing and evaluating, as well as reporting results and progress, are usually undertaken as part of the dynamic character of updating and modifying existing programmes and practices.

Further, these activities are often coordinated across, or engage with, different levels of society and government: international, regional, national, state and local.

## Adaptation and Disaster Risk Reduction

Concepts of, policies for and practices in adaptation, especially at the international level, have merged with two pre-existing discourses, those of disaster risk reduction (DRR) and in a more general way, of development, particularly as sustainable development (Cohen et al. 1998; Birkmann and von Teichman 2010; IPCC 2014; UNISDR 2015; Oliver-Smith et al. 2016; Olsson 2018). For convenience, this volume employs the notion of discourses to stand in for the distinctive associations between the language, practices and institutions associated with these specific concepts. Some official bodies and scholars consider that these separate discourses (disaster risk reduction and sustainable development) should be integrated and not treated as separate in practice and research (IPCC 2014; UNISDR 2015; Oliver-Smith et al. 2016).

Such is the extent of this overlap that the identity of climate change adaptation is becoming difficult to discern, prompting questions such as: Are there significant differences between adaptation and these other discourses? Do these discourses have conflicts or synergies? And, does the presence or absence of any significant discourse boundaries have any political significance? Three inter-related issues are of interest: (1) Determining the extent of overlap, commonality and distinctiveness between these discourses, (2) Functional relationships between climate change adaptation and DRR and (3) Functional relationships between climate change adaptation and development (cf., Birkmann and von Teichman 2010; Pelling 2011; Rivera Escorcia 2016).

DRR covers an array of hazards (both latent natural and human-induced hazards); the climate-related natural hazards are those of mutual interest with climate change adaptation. Traditionally, DRR has been taken as part of the broader development discourse, but with a distinctive identity. Adaptation also addresses that considerable body of climate change impacts that are not considered as climate-related hazards. Climate change adaptation has been seen as having a role to contribute to sustainable development from the outset, but also as arguably having functional interests outside sustainable development (Olsson 2018).

Studies and analyses of climate-related disasters provide concepts, tools and insights into such issues as institutional arrangements and socio-economic impacts relevant to climate change adaptation policy and programmes. In the early days of the climate change discourse, popular imaginations were fuelled by speculations and apocalyptically themed perceptions to make extreme climate events the motif for climate change (Cromwell and Levene 2009; Swyngedouw 2010). Extreme climate-related hazards are of critical importance as, by definition, they exact a great toll on human life, health and welfare and are a major factor in understanding the history and vulnerabilities of many lesser developed nations. These effects can, however, be misleading as representations of climate change impacts. As described above, climate change impacts are far wider and more diverse than their influences on hazardous climate events. There is a risk, therefore, that climate change becomes a scapegoat for every climate-related disaster and used to obviate understanding the socio-economic dimensions of vulnerability and the need to reduce these, thereby undermining DRR efforts.

In practice and research, DRR and disaster responses precedes that of climate change by many years. Following the UNISDR (2015), DRR is achieved by systematically analysing and lowering the risk factors of disasters, involving reducing hazard exposure, lowering vulnerability of people and economic, physical, social and environmental assets, improving preparedness and early warning of hazardous events and 'wise' management of the environment. Under the Sendai Framework, disaster risk reduction is part of sustainable development, so that development must reduce disaster risks and recognising that unsustainable development increases risks and any resulting losses from disasters (UNISDR 2015). Socio-economic factors and politics are taken as the causal factors of disasters, following the mantra that 'there are no natural disasters, only natural hazards' (this perspective supplanting the earlier 'inevitable and unavoidable' approach that focused on disaster recovery, rather than risk reduction) (cf., Kelman 2020).

There are many similarities between climate change adaptation and DRR (see, Box 2.1) (Birkmann and von Teichman 2010; IPCC 2014; UNISDR 2015; Oliver-Smith et al. 2016). In common is that both have developed a set of analytical tools and techniques for identifying vulnerabilities from which response measures can be formulated. An appreciation of poverty is a feature of both discourses, as it is both an outcome of vulnerabilities and a source of vulnerability. Both have also taken up the issues of biodiversity protection and sustainable resource usage, as these play a role in resilience. Also shared are those general barriers to effective policy formulation and implementation. And, somewhat independently, both discourses have taken up the cause of 'mainstreaming' their activities through greater linkage (as discussed below).

What are the differences? Obviously, DRR has a history based in humanitarian relief after events, it deals with present issues and its resilience is based on traditional/indigenous knowledge. Historical records provide the basis for actions in DRR, whereas climate change adaptation is based largely in scientific knowledge, deals with future risks and emerging risks and views traditional and indigenous knowledge as insufficient for resilience in the face of new types and levels of climate hazards (Mercer 2010; Thomalla et al. 2006; Schipper 2009). A further difference is that climate change adaptation deals primarily with gradual and incremental changes, while DRR more typically focuses on emergencies. Institutionally, the discourses have been largely separate, as governments have usually dealt with DRR through civil defence agencies, climate change adaptation through environment and energy agencies. Funding, support, information creation and dissemination systems have been essentially separate. Critically, the involvement of formal political institutions, stakeholders and processes have been largely separated at the sub-national and national levels. In their studies of the practical barriers to linking climate change adaptation and DRR, Birkmann and von Teichman (2010) found these barriers to be mismatches in geographical scale, differences in timescales and the influence of norms.

Despite these differences, the discourses have drawn closer together in recent years. DRR has been adopting preparatory and awareness-raising approaches and shifting towards long-term perspectives; climate change adaptation has been giving greater recognition to the need for community-based approaches and use of traditional/indigenous knowledge. Mercer (2010) views the differences between DRR and climate change adaptation as 'inconsequential for practical application', with these mainly concerning the past-orientation of DRR contrasting with the future-orientation of climate change adaptation. Given that the strategies of DRR and climate change adaptation are so similar, Mercer argues for embedding climate change adaptation in DRR. This partly arises from concerns that climate change adaptation was being defined narrowly and was disinterested in root causes of vulnerability, something that DRR has long been addressing and from the additional funding opportunities that climate change adaptation could bring to DRR.

**Box 2.1 Disasters: Risk and adaptation**

Much of the thinking about adaptation is drawn from the experiences, institutional responses and studies from climate-related disasters. In the popular imagination and fuelled by media and apocalyptically-themed popular entertainment, climate change impacts are made synonymous with 'natural' disasters. Extreme events are of critical importance, but climate change impacts are far wider than affecting the scale, distribution and frequency of dangerous future climatic events. Nonetheless, our thinking about disasters before they occur and what occurs in their aftermath have greatly influenced the agenda, research and practices in climate change adaptation, despite the usual short-term horizon for disaster risk reduction. Key political issues include the social dimensions of risk exposure to severe weather events, as those of socio-economic disadvantage and lesser-developed nations are more vulnerable. Recovery and recuperation following events also evokes questions of the allocation of support and resources across affected communities. Urban settlements, livelihood supports, industries, productive centres, agricultural production are amongst those aspects of social and economic life that can be temporarily disrupted or permanently altered by disasters. Not only do climatic disasters cause immediate social havoc to those involved, but the aftereffects can also cause changes to political institutions and impact political actors. Adaptation measures include capacity-building to address vulnerability to severe weather events, such as low levels of resources, institutional weakness, inadequate knowledge and infrastructure deficiencies. Disruptions can also create opportunities to reconsider political arrangements and for political change when existing authorities lose their legitimacy; i.e., transformational adaptation may be possible.

See, Birkmann and von Teichman (2010), IPCC (2012), and Thomalla et al. (2006).

## Adaptation and Sustainable Development

Illustrative of how the relationship between adaptation and sustainable development have changed through the evolution of climate change, Cohen et al. (1998) described the 'remarkable' circumstances of how the two discourses had virtually ignored each other and had come to represent different understandings of global environmental problems. While some contemporary commentators might endorse such views as still being current, there has clearly been a rapid escalation in the mutual acknowledgement, coordination and overlap between the discourses.

There are clearly good reasons for both discourses to take the other into account (Olsson 2018). For those developing nations overwhelmed by frequent natural hazards, the resources and efforts required for recovery have hampered their longer-term development and capacities to address pressing economic and social problems (cf., Kelman 2020). Climate change poses a particular risk for such places as it increases both the climate-related hazards and associated hazard risks (i.e., disasters undermine development and lesser development tends to create greater risk exposure to natural hazards). Many of the processes contributing to vulnerability relate directly to development metrics, such as poverty and marginalization, conflict and instability, high population growth and high rates of urbanisation, increased habitation of coasts and floodplains and lowering of environmental conditions.

Two different models of the relationship have been offered (Schipper 2007). One is that adaptation activities are effective in reducing vulnerabilities to future impacts so that eventual climate hazards will have greatly reduced impacts, thereby facilitating sustainable development. A second is that sustainable development reduces climate change vulnerability, lowering the risks of climate hazards, thereby becoming a process of adaptation. Even more simply described, the former is an adaptation-led approach to sustainable development and the latter, sustainable development as a form of adaptation. Political interests are expressed at many points in the conjoining of these discourses (Olsson 2018; see, Chapter 3).

## Status Quo or Transformation?

As indicated by the discussion above, adaptation possesses an attribute that makes it attractive to academic scholarship but possibly frustrating to practitioners and the wider public, namely a highly contested and ambiguous identity. Conceptually, it is a difficult idea, with widely varying interpretations within and between academic disciplines (Hulme 2009). As Thomalla et al. (2006) observed, climate hazard reduction has become the subject of four different research communities, namely disaster risk reduction, climate change adaptation, environmental management and poverty reduction. As described above, there has been a tendency to consider adaptation as largely concerned with potential 'natural' disasters; a view that can result in neglecting slow moving and more gradual changes that do not culminate in concentrated and very tangible 'disaster events' but in more diffuse and dispersed harmful, sometimes social and cultural, outcomes (Adger et al. 2013). Several authors argue for the unification of the DRR and adaptation communities (Thomalla et al. 2006; Renaud and Perez 2010; Granberg 2019).

Not only is the idea of adaptation challenging, but all of its major elements contribute to the complexity of understanding, recognizing, formulating and undertaking adaptation. To some extent, these complexities arise from fundamental differences over the purpose or goals of adaptation. In common parlance, adaptation simply aims to alleviate the adverse effects of climate change impacts (Jagers et al. 2009) and also to realise any potential opportunities and benefits (Pelling 2011). A good exemplar of the mainstream understanding of adaptation is those of the IPCC. While the IPCC depict the goals of adaptation as improving the capacity to reduce potential harm from risks, Nelson et al. (2007, p. 397) are more explicit in understanding adaptation goals, to wit:

> ... to maintain the capacity to deal with future change or perturbations to a social-ecological system without undergoing significant changes in function, structural identity, or feedbacks of that system while maintaining the option to develop.

This view focuses on building resilience and since we are interested in the politics of adaptation, an immediately striking feature of this perspective is that adaptation is striving to ensure that the future resembles the present as much as is practicable; in other words, the goal is conservative, focusing on the preservation of the status quo rather than on transition or transformation.

This conservative and one-dimensional view of resilience has been succeeded by notions of 'evolutionary resilience', whereby socio-ecological systems change into new forms following disturbances; i.e., the normal is followed by the 'new normal' (UNISDR 2015). Such changes in perceptions of resilience reflect the broader trends in science away from mechanistic and linear explanations (i.e., Newtonian) and the uptake of complex, non-linear, discontinuous and self-organizing explanations of the natural and social worlds, including socio-ecological systems. Views that changes in ecological systems were directly proportional to external stresses were replaced by recognition that 'tipping points' could be reached through minor stimuli—or conversely, that systems could be robust in the face of major influences (Lenton 2011). At the macro scale, these tipping points could occur in natural systems, such as in the Arctic or the Antarctic ice sheets or the Amazon rainforest, potentially leading to large-scale irreversible change. Lately this idea has widened to include impacts of ecological change on social systems as well and how this can influence adaptive capacity (Garschagen and Solecki 2017; Fünfgeld 2017).

Pelling describes socially constructed 'thresholds' that are related to what is considered acceptable risk by social actors and, when crossed, can result in violent disruption of equilibriums, forcing systems to enter new states (2011, pp. 45–47). High impact events can also function as social tipping points or thresholds as they can lead to new development trajectories that can break up inertia and path dependency maintaining existing policies and systems (Rosenzweig and Solecki 2014). Such changes need not result from external stressors but can also occur from internal causes during apparent steady-state conditions; i.e., social systems can also have 'chaotic' characteristics. In the same way, shifts in social, economic and political conditions can impact vulnerabilities and adaptive capacity in positive or negative ways (Garschagen and Solecki 2017).

Socio-ecological systems can be depicted as following ecological cycles by developing in phases (variously described), but basically cycle through periods of growth, decline/collapse, and then reorganization. As Davoudi (2012) wondered, if socio-ecological systems behave in this manner and our understandings based on history are without predictive capacities, does this make planning impossible? In this regard, the concept of resilience is of particular interest, having as it does a high and growing profile in adaptation discourse and now serving as a mainstream goal of adaptation (Woodruff et al. 2018).

At its simplest, resilience concerns protecting—from a stakeholder's perspective—those things threatening social welfare and managed ecological systems from current/future climate change impacts. Pelling stated that (2011, p. 55):

> … resilience is a form that seeks to secure the continuation of desired systems functions into the future in the face of changing context, through enabling alteration in institutions and organisational form.

This is achieved '… through social learning and self-organization to enable technological evolution, new information exchange or decision-making procedures' (Pelling 2011, p. 56). Additionally, Pelling states that resilience, and the process of resilience building, may require changes in values and institutions within managing organisations.

Contemporary usage of the resilience concept is rooted in applying an engineering concept to ecology by Holling (1973) for describing the extent of disruption that an ecological system can absorb before its structure is altered, with its subsequent uptake in studies of socio-ecological systems. As ecological systems are obviously not mechanical, some accommodations were necessary. A feature of ecological systems is multiple points of equilibria; a disturbed ecological system can achieve stability in a new state, whereas engineered systems have a single point of stability. Nonetheless, such conceptual understandings of resilience concerned the recovery of stable systems to a condition of equilibria ('balanced Nature'). Indeed, the ideas of resilience can be found throughout the canon of normative social sciences, health and medicine, and other fields and usually carrying assumptions of the desirability of equilibrium, ranging from psychological recovery from trauma to urban recovery from disasters.

Not unexpectedly, anthropomorphizing the concept of resilience from ecology to socio-ecological systems provoked criticism that encompasses the following arguments (Cote and Nightingale 2012; Davoudi 2012; Brown 2013):

- Human agency, that which essentially defines the social, is directed at shaping and directing the socio-ecological relationships, which runs counter to the determinism of adaptive cycles in natural systems
- Social relations, notably that of politics, play an important role in devising and enabling resilience building and in shaping the distribution of the resulting costs and benefits
- Social adaptation has a purposefulness—the answer to the question 'what is resilience for?—that cannot be readily ascribed to natural systems, and

- Defining the boundary of social systems poses particular difficulties; and their definition influences resilience responses, such as over the inclusion/exclusion of social groups, localities and time horizons, but also questions of epistemology, such as the selection and engagement of scholarly disciplines.

Much in the manner of the concept of sustainability, the transition from the natural sciences to wider applications has seen the concept of resilience move from a relatively narrow meaning to one of wider application. Almost invariably, however, the meaning of resilience has become marked by confusion, difficulties in application and measurement, and the risk that it will become 'nothing more than a rhetorical device' (Davidson et al. 2016). Set against this problem is the recognition of adaptation studies and practices taking up a political agenda—as described above by Pelling (2011), namely that adaptation be transformational (see, Box 2.2).

**Box 2.2 Pelling on adaptation frameworks**

Pelling (2011) identifies three contrasting frameworks or pathways for adaptation practices, the first of which—resilience—is that most commonly advanced.

**Resilience**: Adaptation measures designed for preventing change harmful to existing functions of systems through changes to institutions and organisations. Resilient outcomes feature 'social systems' with 'functional persistence', 'self-organisation' and 'social learning'.

**Transition**: Adaptation measures promoting incremental change to in political and cultural institutions; these are not intended to, or result in, 'regime change'.

**Transformation**: Adaptation measures that reform existing social contracts for progressive social goals; this involves addressing the root causes of vulnerability.

Adaptation has a number of implications as expressed in the contrasting pathways for adaptation and the extent to which adaptation is engaged in other policy and practice realms (Eriksen et al. 2015). Much of scholarship and debate over adaptation has centred on the concept of resilience, having as it does a high and growing profile in many fields, including in international relations, disasters, human development and planning discourses (Davoudi 2012; Cote and Nightingale 2012; Brown 2013; Woodruff et al. 2018). Its meaning, and the meaning of associated concepts, varies between the fields in which it is used. Views as to its meaning are scarcely less varied than those on adaptation. Making matters less clear from a scholarly perspective, the large scholarly literature on resilience now overlaps and intersects with that of sustainability, that behemoth of contested concepts in environmentalism. There is evidence to suggest that much of the agenda presented under the rubric of 'sustainability' in earlier decades now appears as 'resilience'. Nonetheless, the concept of resilience building and the concept of resilience itself is central to the adaptation debate, along with the emerging political considerations, as discussed in the following chapter.

## References

Adger, W. N., Barnett, J., Brown, K., Marshall, N., & O'Brien, K. (2013). Cultural dimensions of climate change impacts and adaptation. *Nature Climate Change, 3,* 112–117.

Adger, W. N., Dessai, S., Goulden, M., Hulme, M., Lorenzoni, I., Nelson, D., et al. (2009). Are there social limits to adaptation? *Climatic Change, 93,* 335–354.

Ayers, J. M., & Huq, S. (2008). The value of linking mitigation and adaptation: A case study of Bangladesh. *Environmental Management, 43*(5), 753–764.

Barnett, J., & O'Neil, S. J. (2010). Maladaptation. *Global Environmental Change, 20*(2), 211–213.

Bassett, T. J., & Fogelman, C. (2013). Déjà vu or something new? The adaptation concept in the climate change literature. *Geoforum, 48,* 42–53.

Bauer, A., Feichtinger, J., & Steurer, R. (2012). The governance of climate change adaptation in 10 OECD countries: Challenges and approaches. *Journal of Environmental Planning Policy & Management, 14*(3), 297–304.

Berrang-Ford, L., Ford, J. D., & Paterson, J. (2011). Are we adapting to climate change? *Global Environmental Change, 21,* 25–33.

Biagini, B., Bierbaum, R., Stults, M., Dobardzik, S., & McNeeley, S. M. (2014). A typology of adaptation actions: A global look at climate adaptation actions financed through the Global Environment Facility. *Global Environmental Change, 25*, 97–108.

Birkmann, J., & von Teichman, K. (2010). Integrating disaster risk reduction and climate change adaptation: Key challenges—scales, knowledge, and norms. *Sustainability Science, 5*(2), 171–184.

Boezeman, D. (2015). *Authoritative knowledge for climate change governance*. Nijmegen: Radboud University Nijmegen.

Brown, K. (2013). Global environmental change I: A social turn for resilience? *Progress in Human Geography, 38*(1), 107–117.

Burkhard, B., & Maes, J. (Eds.). (2017). *Mapping ecosystem services*. Sofia: Pensoft Publishers.

Cohen, S., Demeritt, D., Robinson, J., & Rothman, D. (1998). Climate change and sustainable development: Towards dialogue. *Global Environmental Change, 8*(4), 341–371.

Cote, M., & Nightingale, A. J. (2012). Resilience thinking meets social theory: Situating social change in socio-ecological systems (SES) research. *Progress in Human Geography, 36*(4), 475–489.

Cromwell, D., & Levene, M. (Eds.). (2009). *Surviving climate change: The struggle to avert global catastrophe*. London: Pluto Press.

Davidson, J. L., Jacobson, C., Lyth, A., Dedekorkut-Howes, A., Baldwin, C. L., Ellison, J. C., et al. (2016). Interrogating resilience: Toward a typology to improve its operationalization. *Ecology and Society, 21*(2), 27. https://doi.org/10.5751/ES-08450-210227.

Davoudi, S. (2012). Resilience: A bridging concept or a dead end? *Planning Theory & Practice, 13*(2), 299–307.

Elmquist, T., Gatzweiler, F., Lindgren, E., & Liu, J. (2019). Resilience management for healthy cities in a changing climate. In M. R. Marselle, J. Stadler, H. Korn, K. N. Irvine, & A. Bonn (Eds.), *Biodiversity and health in the face of climate change* (pp. 411–424). Springer Nature: Cham.

Eriksen, S. E., Nightingale, A. J., & Eakin, H. (2015). Reframing adaptation: The political nature of climate change adaptation. *Global Environmental Change, 35*, 523–533.

Fankhauser, S., Smith, J. B., & Tol, R. S. J. (1999). Weathering climate change: Some simple rules to guide adaptation decisions. *Ecological Economics, 30*(1), 67–78.

Fünfgeld, H. (2015). Facilitating local climate change adaptation through transnational municipal networks. *Current Opinion in Environmental Sustainability, 12*, 67–73.

Fünfgeld, H. (2017). Institutional tipping points in organizational climate change adaptation processes. *Journal of Extreme Events, 4*(1), 1750002. https://doi.org/10.1142/S2345737617500026.

Garschagen, M., & Solecki, W. (2017). Tipping points in adaptive capacity and adaptation processes. *Journal of Extreme Events, 4*(1), 1702002. https://doi.org/10.1142/S234573761702002X.

Granberg, M. (Ed.). (2019). *Göteborgsregionen och klimatrisker* (Rapport 2019:1). Karlstad: The Centre for Climate and Safety, Karlstad University.

Granberg, M., & Glover, L. (2014). Adaptation and maladaptation in Australian national climate change policy. *Journal of Environmental Policy & Planning, 16*(2), 147–159.

Holling, C. S. (1973). Resilience and stability of ecological systems. *Annual Review of Ecology and Systematics, 4,* 1–23.

Hulme, M. (2009). *Why we disagree about climate change: Understanding controversy, inaction and opportunity*. Cambridge: Cambridge University Press.

Hulme, M., Barrow, E. M., Arnell, N. W., Harrison, P. A., Johns, T. C., & Downing, T. E. (1999). Relative impacts of human-induced climate change and natural climate variability. *Nature, 397*(25), 688–691.

IPCC. (1990). *Climate change 1990: The IPCC impacts assessment*. Canberra: Australian Government Publishing Service.

IPCC. (1996). *Climate change 1995: Impacts, adaptations, and mitigation of climate change: Scientific-technical analyses*. Cambridge: Cambridge University Press.

IPCC. (2001). *Climate change 2001: Impacts, adaptation and vulnerability*. Cambridge: Cambridge University Press.

IPCC. (2007). *Climate change 2007: Impacts, adaptation and vulnerability*. Cambridge: Cambridge University Press.

IPCC (Intergovernmental Panel on Climate Change). (2012). *Managing the risks of extreme events and disasters to advance climate change adaptation*. Cambridge: Cambridge University Press.

IPCC. (2014). *Climate change 2014: Impacts, adaptation, and vulnerability*. Cambridge: Cambridge University Press.

Jabareen, Y. (2013). Planning the resilient city: Concepts and strategies for coping with climate change and environmental risk. *Cities, 31,* 220–229.

Jagers, S., Duus-Otterström, G., & Stripple, J. (2009). Fair and feasible climate change adaptation. *Statsvetenskaplig Tidskrift, 111*(1), 53–58.

Juhola, S., Linnér, B.-O., Glaas, E., & Neset, T.-S. (2016). Redefining maladaptation. *Environmental Science & Policy, 55,* 135–140.

Kahn, M. E. (2016). The climate change adaptation literature. *Review of Environmental Economics and Policy, 10*(1), 166–178.

Kelman, I. (2020). *Disaster by choice: How our actions turn natural hazards into catastrophes*. Oxford: Oxford University Press.

Lenton, T. M. (2011). Early warning of climate tipping points. *Nature Climate Change, 1,* 201–209.

Magnan, A., Schipper, E. L. F., Burkett, M., Bharwani, S., Burton, I., Eriksen, S., et al. (2016). Addressing the risk of maladaptation to climate change. *Wiley Interdisciplinary Reviews: Climate Change, 7*(5), 646–665.

Mercer, J. (2010). Disaster risk reduction or climate change adaptation: Are we reinventing the wheel? *Journal of International Development, 22,* 247–264.

Moloney, S., Fünfgeld, H., & Granberg, M. (2018). Climate change responses the global to local scale: An overview. In S. Moloney, H. Fünfgeld, & M. Granberg (Eds.), *Local action on climate change: Opportunities and constraints* (pp. 1–16). London and New York: Routledge.

Moser, S. C., & Ekstrom, J. A. (2010). A framework to diagnose barriers to climate change adaptation. *Proceedings of the National Academy of Sciences of the United States of America (PNAS), 107*(51), 22026–22031.

Nelson, D., Adger, W. N., & Brown, K. (2007). Adaptation to environmental change: Contributions of a resilience framework. *Annual Review of Environment and Resources, 32,* 395.

Neset, T.-S., Asplund, T., Käykhö, J., & Juhola, S. (2019). Making sense of maladaptation: Nordic agriculture stakeholders' perspectives. *Climatic Change, 153,* 107–121. https://doi.org/10.1007/s10584-019-02391-z.

Nightingale, A. J., Eriksen, S., Taylor, M., Forsythe, T., Pelling, M., Newsham, A., et al. (2019). Beyond technical fixes: Climate solutions and the great derangement. *Climate and Development*. https://doi.org/10.1080/17565529.2019.1624495.

Nyong, A., Adesina, F., & Osman Elasha, B. (2007). The value of indigenous knowledge in climate change mitigation and adaptation strategies in the African Sahel. *Mitigating and Adaptation Strategies for Global Change, 12*(50), 787–797.

Oliver-Smith, A., Alcántara-Ayala, I., Burton, I., & Lavell, A. M. (2016). *Forensic investigations of disasters (FORIN): A conceptual framework and guide to research* (IRDR FORIN Publication No. 2). Beijing: The Integrated Research on Disaster Risk Programme (IRDR).

Olsson, D. (2018). *Conditions of 'sustainability': The case of climate change adaptation in Sweden*. Karlstad: Karlstad University Press.

Oreskes, N. (2005). The scientific consensus on climate change. *Science, 306*(5702), 1686. https://doi.org/10.1126/science.1103618.

Paavola, J., & Adger, W. N. (2006). Fair adaptation to climate change. *Ecological Economics, 56*(4), 594–609.

Palutikof, J. P., Leitch, A. M., Rissik, D., Boulter, S. L., Campbell, M. J., Perez Vidaurre, A. C., et al. (2019). Overcoming knowledge barriers to adaptation using a decision support framework. *Climatic Change, 153,* 607–624.

Parry, M., Arnell, N., Berry, P., Dodman, D., Fankhauser, S., Hope, C., et al. (2009). *Assessing the costs of adaptation to climate change: A review of the*

*UNFCCC and other recent estimates*. London: International Institute for Environment and Development and Grantham Institute for Climate Change.

Patterson, J. J., & Huitema, D. (2019). Institutional innovation in urban governance: The case of climate change adaptation. *Journal of Environmental Planning and Management, 62*(3), 374–398.

Pelling, M. (2011). *Adaptation to climate change: From resilience to transformation*. London: Routledge.

Pielke, R. A., Prins, G., Rayner, S., & Sarewitz, D. (2007). Lifting the taboo on adaptation. *Nature, 445*(8), 597–598.

Renaud, F., & Perez, R. (2010). Climate change vulnerability and adaptation assessments. *Sustainability Science, 5,* 155–157.

Rivera Escorcia, C. M. (2016). *Disaster risk management and climate change adaptation in urban contexts: Integration and challenges*. Lund: Lund University.

Rosenzweig, C., Karoly, D., Vicarelli, M., Neofotis, P., Wu, Q., Casassa, G., et al. (2008). Attributing physical and biological impacts to anthropogenic climate change. *Nature, 453,* 353–357.

Rosenzweig, C., & Solecki, W. (2014). Hurricane Sandy and adaptation pathways in New York: Lessons from a first-responder city. *Global Environmental Change, 28*, 395–408.

Schipper, E. L. F. (2006). Conceptual history of adaptation in the UNFCCC process. *Review of European Community and International Environmental Law, 15*(1), 82–92.

Schipper, E. L. F. (2007). *Climate change adaptation and development: Exploring the linkages* (Working Paper 107). Norwich: Tyndall Centre for Climate Change Research.

Schipper, E. L. F. (2009). Meeting at the crossroads? Exploring the linkages between climate change adaptation and disaster risk reduction. *Climate and Development, 1,* 16–30.

Schipper, E. L. F., & Burton, I. (Eds.). (2009). *The Earthscan reader in adaptation to climate change*. London: Earthscan.

Strauss, S. (2015). Climate adaptation: Cultural knowledge and local risks. *Nature Climate Change, 5,* 624–625.

Swyngedouw, E. (2010). Apocalypse forever? Post-political populism and the spectre of climate change. *Theory, Culture and Society, 27*(2–3), 213–232.

Tembya, O., Sandall, J., Cooksey, R., & Hickey, G. M. (2016). How do civil servants view the importance of collaboration and scientific knowledge for climate change adaptation? *Australasian Journal of Environmental Management, 23*(1), 5–20.

Thomalla, F., Downing, T., Spanger-Siegfried, E., Han, G., & Rockström, J. (2006). Reducing hazard vulnerability: Towards a common approach between disaster risk reduction and climate adaptation. *Disasters, 30*(1), 39–48.

UNISDR. (2015). *Sendai framework for disaster risk reduction 2015–2030*. Geneva: UNISDR.

Verchot, L. V., Van Noordwijk, M., Kandji, S., Tomich, T., Ong, C., Albrecht, A., et al. (2007). Climate change: Linking adaptation and mitigation through forestry. *Mitigating and Adaptation Strategies for Global Change, 12*(5), 901–918.

Woodruff, S. C., Meerow, S., Stults, M., & Wilkins, C. (2018). Adaptation to resilience planning: Alternative pathways to prepare for climate change. *Journal of Planning Education and Research*, 1–12. https://doi.org/10.1177/0739456X18801057.

CHAPTER 3

# Emerging Political Considerations in Climate Change Adaptation

**Abstract** This chapter provides an overview of key influences on the development of adaptation (both conceptually and as a practice) that entail the political. It begins with an account of the interplay of different political and associated elements that are shaping the development of adaptation in what is described as a 'complex arena'. Adaptation politics are taken as political values, processes, interests and outcomes involving states, civil society and corporations. Following this, we examine the central place of scientific knowledge in the adaptation discourse and the critiques of, and challenges against, its apparent hegemony. Neoliberalism is recognised as a major socio-economic force in global politics and its influence on adaptation is examined. Indigenous peoples are now receiving greater attention in adaptation and this is another emerging political consideration shaping policy, practices and programmes, including the recognition of traditional and indigenous knowledge and the calls for community-based adaptation. Commons are also described in this context. Closing the chapter is a discussion of the role of the state in adaptation, including the essential reasons for its involvement and the implications therein.

**Keywords** Policymaking · Neoliberalism · Planning · Scientific knowledge · Technocracy · Traditional knowledge



L. Glover and M. Granberg,
*The Politics of Adapting to Climate Change*,
https://doi.org/10.1007/978-3-030-46205-5_3

## A Complex Arena

Adaptation politics are no less narrow than the field of adaptation itself (Swyngedouw 2010a, b; Nightingale 2017). Taking the simple step-wise approach to adaptation planning of assessment, planning and design, implementation, and monitoring and reporting, politics can be readily seen to be engaged at every stage. Adaptation planning occurs in the many spheres of society (state, corporate and civil society) and ranges across the realms of different levels of government, households, international agencies, NGOs, private enterprise, and communities and other social groups, and occurs in cooperative arrangements between these entities. Each of these spheres expresses its own set of political attributes and influences. There are also key 'framing' political decisions that occur before the planning processes begin, such as selecting issues to address, choosing locations for action, identifying the communities involved and so on (Adger et al. 2005; Unwin and Jordan 2008; Granberg and Glover 2014; Moloney and Fünfgeld 2015; Moloney et al. 2018b; Granberg and Nyberg 2018). Adger (2003), for example, viewed adaptation as a collective action problem with implications for public- and private-sector roles in developing necessary social capital. Outside such framing are the contextual political factors, such as the influences of culture, economics, geography, history and social relations. As discussed above, the political can be identified in such aspects as the influence of political values, use of power and influence, the formation and functioning of institutions, and in creating and legitimating knowledge. Although politics might be found wherever it is sought in the adaptation discourse and that there are many aspects yet to be examined in depth, the recognition of politics in the discourse has resulted from activity at a number of specific 'pressure points'.

Several underlying trends are responsible for the growing awareness of, interest in and scholarship about politics, which has a bearing on such matters as stakeholder involvement, knowledge legitimation and decision making processes. Adaptation politics have come to reflect a broader debate in progressive politics, that of redistribution and recognition, with the concomitant rise of 'identity politics' (see, Box 3.1). This is significant in several ways: it reveals the dynamic character of adaptation politics and shows how deeper political developments and the role of social movements have come to influence adaptation politics, leading to the observation that adaptation politics are not a separate sphere of activity, but one linked, reflexively, to wider political developments. In one sense, adaptation is another, albeit distinct, theatre for the performance of contemporary politics. One such trend is that the field of climate change has 'caught up' with the general interest in 'traditional ecological knowledge' (TEK) and 'indigenous traditional knowledge' (ITK) that many fields and disciplines active in global environmental change issues have been studying for many years (Berkes 1999; Dekens 2007; Houde 2007; Leonard et al. 2013). Here, the field of adaptation has responded and evolved, with official recognition of the importance of TEK and ITK.

**Box 3.1 Progressive politics: Redistribution and recognition**

For the entire span of progressive politics of the industrial era, the Holy Grail of social justice has been material equality, with the greatest failing of industrial capitalism viewed as its perpetuating and exacerbating of socio-economic inequalities. Uprisings, revolutions and wars have been rationalised and legitimated on the cause of material redistribution under the ideal of material equity, if not equality. At the more reformist and incremental end of the political spectrum, advocacy of redistribution has also underlain the growth of the welfare state and arguably the Keynesian model of state-managed economic development, featuring such policies as corporate tax, minimum wages, pension schemes, progressive income tax and social security. Redistribution (of wealth, income and opportunity) is also a marker for different political ideologies, being endorsed by socialism and rejected by liberalism and conservatism. Redistribution advocates pursue material equality for its expression of fairness; i.e., it is a just cause. Equity ensures that individual's needs are met and can be fulfilled, i.e., it enables self-actualisation. Inequality, to this group, is an artefact of society (not Nature). Equality before the law and administration alone is insufficient to overcome pre-existing inequalities; material equality requires more direct measures of redistribution. More recently, progressive politics has become concerned with the political cause of representation (aka 'identity politics'), with some seeking to put recognition at the centre of social justice. Largely, this is a cause directed at satisfying the needs of minorities subject to injustices by majorities without making assimilation the price of acceptance; in some cases, there is a dominant cultural group or identity rather than a majority. Social justice is sought through the recognition of, and respect for, social and cultural difference within society. While political freedom is the goal of identity politics, it is usually not organised around formal political parties or formal belief systems. In some respects, this movement represents a 'cultural turn' for progressive politics. Key concepts involved include autonomy, difference, interdependence and self-determination. Within pluralistic and multicultural societies, a degree of formal recognition of equal rights in terms of the state are readily secured, often as enshrined in national Constitutions. However, such matters as freedom of expression can easily run counter to the interests of groups with opposing views; one person's freely expressed opinion can readily be another's hate speech. Further, the making of identity can be highly problematic; identity is often defined as differences between 'others', identity can be multifaceted and flexible rather than singular, and can alter according to context. This development in progressive politics has not been without problems, with concerns that it is being pulled apart by competing priorities. Some consider that the two priorities are two sides of the same coin and should be synthesized or at the least, pursued cooperatively.

See, Hobsbawm (1996), Moya (2000), and Fraser (2001).

Indigenous peoples have lengthy experience with accommodating to climatic change; their experiences in coping with, and adapting to, climate change has accumulated as traditional knowledge and is passed between successive generations (e.g., Nakashima et al. 2012; Makondo and Thomas 2018; Nyong et al. 2007). Traditional knowledge features close observation of biophysical conditions, local-scale knowledge and involves knowledge-sharing systems (Trundle 2018). TEK can be viewed as a knowledge-practice-belief system (Berkes 1999), providing factual observation, knowledge of management systems, and an ethical, cultural and identity construct. Such climate knowledge is embedded within a cultural context and also comprises an integrative component of culture.

Modern programmes for adaptation have been largely based on western and scientific knowledge, placing them directly at odds with these older knowledge forms and local cultural contexts. Traditional knowledge has usually been omitted or downplayed in these endeavours, thereby marginalising indigenous and traditional societies in this regard. More recently, adaptation programmes have been incorporating traditional knowledge. It is recognised in the UNFCCC Paris Agreement (2016, Article 7.5) and the earlier Cancun Adaptation Framework (2010, para. 12). There have a been number of studies into integrating scientific and traditional knowledge. Benefits of using indigenous and traditional knowledge include identifying options and actions that are culturally appropriate, that match local socio-ecological circumstances and that contribute to social cohesion. Applying such knowledge narrowly entails the risk of overlooking contemporary and external economic and other factors that are contributing to community and household vulnerabilities. These changes mark the efforts to consider cultural aspects of vulnerability and adaptation responses by governance bodies, such as the case made by Adger et al. (2013). Psychological and sociological factors have been brought into adaptation scholarship as part of this change (Grothmann and Patt 2005; Gifford 2011).

Associated with this change has been the rise in community-based responses, which promote community participation and engagement with adaptation measures and programmes, often promoting the role for indigenous and traditional knowledge. Community-based approaches to adaptation are largely drawn from development studies, policies and practices (Tompkins and Adger 2004; Chu et al. 2016). Particular attention has been directed towards the most vulnerable communities in developing nations. For the most part, community-based adaptation is a process of engagement by government agencies and international development bodies. A stronger (and rarer) form is adaptation from community-led processes; its promoters believe that this better reflects community knowledge, priorities, and interests, thereby resulting in improved planning and preparation. In many respects, the practice of community-based adaptation has been challenging, as it requires building trust with local communities, communicating the risks of climate change in accessible ways, understanding local vulnerabilities and then developing appropriate actions with communities. Ecosystem-based adaptation has been joined with community-based approaches to form 'Community-based Natural resource management' (Fabricius and Koch 2004; del Mar Delgado-Serrano et al. 2018).

Adding other forms of knowledge to scientific knowledge, taking up ecosystem-based adaptation models and including community-based approaches in planning and implementation have all been associated with the conjoining of adaptation and development activities by states, international agencies and NGOs (Birkmann and von Teichman 2010; Bosomworth and Gaillard 2019; Paterson and Charles 2019). Largely, this has been a mutual adoption by both camps as promoted by international agencies, activists and scholars. Adaptation has been added to many development programmes and a number of adaptation programmes have established links to development programmes. A few aspects of these overlapping interests have stood out; one is a mutual interest in questions of poverty and disadvantage in lesser developed nations. Another overlap is that many development programmes have come to include 'natural' disasters and disaster risk in their range of concerns; as so many of these risks and events are climate related, this brought development programs into the ambit of adaptation programmes. A further overlap is that the preparatory and planning aspects of adaptation fits with many aspects of economic development and sustainable development planning. Linking these overlapping themes and interests is the larger cause of sustainable development and in recent years, as the scale and scope of adaptation has expanded, it has become more engaged with sustainable development goals, concepts, programmes and practices. There are still distinct core aspects to the fields of sustainable development, disaster and adaptation, but now there is considerable overlap between them and engagement with these other fields has been a major factor in the evolution of climate change adaptation research, studies, concepts, plans and programmes. Additional to the linkage to the development discourse is that to the existing urban and land planning discourse, as noted above.

Adding new actors, stakeholders and forms of knowledge drives the need for new adaptation concepts, ideas, institutions and practices. These political developments are also driven by changes in relationship to other discourses, most importantly to that of development going from acknowledgement and recognition to formal integration into institutional practice. Another change has occurred due to what Füssel and Klein (2006) describe as an evolution in vulnerability assessments, including the trend from being science-driven to being policy-driven, which entails a shift towards the realm of politics, based in the observation that adaptation entails the transfer of resources (as funding, technologies and expertise) (see also, Keulertz et al. 2018). As stated in Chapter 1, an interest in institutional aspects of adaptation promotes awareness and interest in political practices and outcomes. A central trend in climate change adaptation policy and plans is the decentralisation of adaptation responses, with a growing array and range of local activity, notably by local government (e.g., Storbjörk 2007; Wilson 2007; Agrawal and Perrin 2008; Measham et al. 2011; Moloney et al. 2018a; Granberg et al. 2019). A widely-held view is that the bulk of climate change's social impacts will be experienced at the local scale (i.e., 'adaptation is local'), thereby reinforcing the case for place-based adaptation responses (Farber 2013; Sovacool and Brown 2017; Moloney et al. 2018a; Granberg et al. 2019). Local governments, local NGOs, local enterprises and communities are in the 'front line' of adaptation responses in many instances. Building and enhancing adaptation capabilities has been increasingly seen as requiring suitably design, resourced and equipped institutions at every relevant sphere of government (e.g., Pelling 2011; Marshall 2015). Agrawal et al. (2012), reviewed projects in 47 least developed nations and recommended that national planners support local efforts by increasing local institutional capacities, empowering local communities, increasing information sharing and improving accountability of local decision makers to their communities.

Maturation of the adaptation discourse and the challenge of policy implementation is requiring scholars and practitioners to consider a range of institutional, social and cultural issues: legal and regulatory frameworks, policy instruments, administrative procedures and coordination between sectors and actors and social 'opportunities and constraints' (Moloney et al. 2018a). Scale is another dimension of the complexity of adaptation politics (see, Box 3.2) (Granberg et al. 2019). Adaptation scholars have employed a variety of approaches for considering these dimensions of adaptation concepts, policy and action. This has drawn the discourse towards the political, albeit without necessarily explicitly engaging with political aspects and ideology.

**Box 3.2 Adaptation politics and scale**

Scale is a confounding but critical aspect of climate change adaptation. As different scales are considered (cf., local, regional, catchment, national or global), the context of adaptation alters greatly administratively, geographically, politically and in a myriad of other ways. Two issues stand out relating to causality in relation to vulnerability assessment and to the characteristics of response measures as transformative socio-ecological system change. Firstly, it is frequently stated that adaptation is primarily response to local conditions because this is the locus of individual, household and community vulnerabilities. While the expression of vulnerability may be acute locally, the proximate causes may well be at higher orders of scale. Poverty is a good exemplar of this asymmetry; its local expression adds to vulnerability, but its causes are most likely to result from higher-order socio-economic factors. Such reasoning lies behind the claim, problematic that it might be, that 'the best form of adaptation is economic development'. Secondly, transformative socio-ecological change is understood as higher-order change. Many of analogies and conceptual tools describing transformative change come from the natural sciences' insights into higher-order change, especially evolutionary theories. A key component of these theories is explaining how local adaptations (i.e., lower-order scales) become agents for change at higher-order levels (i.e., system transformation). Both these issues complicate political analysis as climate change adaptation has aspects of being a multi-scalar problem.

A manifestation of the growing interest in the political aspects of adaptation has been the promotion of 'values' and the need for their incorporation into vulnerability assessment and in adaptation policy and planning through so-called 'values-based approaches' (Adger et al. 2009; O'Brien and Wolf 2010; Eriksen et al. 2015). O'Brien and Wolf (2010) also argue for a values-based approach to adaptation, noting that adaptation has political implications as hierarchies and value-prioritisation are involved in decision making. For others, adaptation responses occur within a pre-existing political context (e.g., McEvoy and Fünfgeld 2013). For Pelling (2011, p. 5), no aspect of context is more important than power:

> Power lies at the heart ... of adaptation. Power asymmetries determine for whom, where and when the impacts of climate change are felt and the scope for recovery. The power held by an actor in a social system, translated into a stake for upholding the status quo, also plays a great role in shaping an actor's support or resistance towards adaptation or the building of adaptive capacity when this has implications for change in social, economic, cultural or political relations, or in the ways natural assets are viewed and used. Accepting that adaptation is contested makes interpreting adaptation as progressive hostage to the observer's viewpoint.

As the challenge of climate change adaptation is evolving into a mainstream public policy activity, governments and official bodies around the world are exploring potential impacts, informing communities of risks and making policies and plans for adapting to future climate change.

Pelling (2011) cautions, it should not be taken that the developments in adaptation policies, plans and programs are all steps forward on a progressive political agenda. Many of these developments may reinforce existing social-economic inequalities or contribute to new forms and distributions of such inequalities. For example, with the routine incorporation of adaptation into existing development policy and planning policy, critics have noted that this has entailed uncritically adopting what Manuel-Navarrete et al. (2009, p. 2) describe as '... the hegemony of particular development visions and governance structures.' A coalescence of factors promoting an interest in the political as a consequence of the approach and governance of adaptation can be identified. On top of the existing socio-economic, environmental and other circumstances are the political interests expressed by the development and planning discourses, to which can be added the impacts on these factors by climate change. Therefore, climate change adaptation decisions are deeply political, influence social relations, involve resource and power redistribution and '... are intrinsically bound to how people and policymakers interpret adaptation and the unspoken assumptions and goals they have in mind' (Remling 2018, p. 478). Other critics have claimed that the scientific and technocratic perspective of adaptation governance has led to de-politicisation of adaptation (Schulz and Siriwardane 2015). It has also been stated that neoliberalism contributes to this de-politicisation through processes that '... forecloses dissent, conflict and the possibility of a different future' (Swyngedouw 2010a, p. 228). Below, the volume will address these phenomena, starting with science and followed by a discussion on neoliberalism and adaptation.

## Does Adaptation 'Belong' to Science?

Examining the 2014 IPCC assessment reveals the 'difficulty' of politics for adaptation governance, as the report neither directly canvasses politics nor finds a place for it in reviewing the state of climate change knowledge. When devising categories of adaptation options, the IPCC considers structural/physical, social (as educational, informational and behavioural forms) and institutions (i.e., economic, laws and regulations, and government policies and programs forms), nearly all of which impinge on political interests, but politics are not reviewed explicitly (see, IPCC 2014, ch. 14). IPCC consideration of the limits, constraints and opportunities to adaptation draws on a relatively small number of political studies and examines ethical issues, but seemingly eschews addressing politics directly (see IPCC 2014, ch. 16).

There is nothing remarkable in the IPCC's technical orientation to adaptation policy. By-and-large, governments and international agencies have approached adaptation as an exercise of centralized planning, drawing on science and technocratic expert advice (Head 2014; Eriksen et al. 2015). Yet, it is in the IPCC's coverage of adaptation economics (see IPCC 2014, ch. 17) that the reduction of political values and interests to economic concepts and expressions is most evident and, by implication, the most sublimated in the treatment of such items as cost-benefit analysis, decision making and cost of adaptation. To underscore this point, the 2014 report considers social science themes such as culture, economics, ethics, governance, innovation and risk, all of which can be reasonably viewed as posing challenges of definition, having contested meanings and complex interactions with other fields, yet politics stands out as the field of knowledge and activity that the IPCC, at the very least, does not consider critical in advising policymakers charged with responsibilities for adaptation policy. By avoiding raising politics in adaptation, this approach to adaptation governance presents itself as a neutral and apolitical stakeholder in a discourse absent of politics.

Thus, adaptation measures in government institutions for local and national policy, planning and practice, and international development initiatives have been described variously as 'rational', 'objective' and/or 'linear-technocratic', subject to 'scientisation', having 'a self-evident analytical framework and normative goal' and overall as being apolitical or de-politicised (Swyngedouw 2010a; b). Dewulf discerns the differences between socio-political and socio-technical adaptation responses as rooted in framing (Dewulf 2013, p. 324):

> While the technoscientific perspective focuses on aspects like decision trees, impact assessments and technological solutions, the sociopolitical perspective directs the attention toward the high stakes, uncertainties, communication strategies, power relations, and equity issues related to climate change adaptation.

Also, the scientific framings of vulnerability in earlier studies of climate change vulnerability meant that 'human-security framings' were ignored or substantially downplayed (Füssel and Klein 2006; O'Brien et al. 2007). Adger et al. have depicted the goals of governance institutions involved in adaptation thus (2013, p. 115):

> Government action seeks to allocate resources efficiently to affect a desirable distribution in areas where autonomous actions and markets fail to do so.

A central trend in climate change adaptation policy and plans is the decentralisation of adaptation responses, with a growing array and range of local activity, notably by local government (e.g., Storbjörk 2007; Wilson 2007; Agrawal and Perrin 2008; Measham et al. 2011; Moloney et al. 2018a).

Another motivation for recognizing political factors arose from views that adaptation practices had implications for development issues, particularly negative effects for developing nations, regions and communities. Barnett (2010) has identified the limitations of the first generation of impacts assessments (on which the bulk of adaptation studies is based), namely that these studies examined the effects of climate system changes on biophysical systems and then subsequently on social systems, downplaying issues of social vulnerability rendering them secondary and 'subordinate'. Barnett stated that (2010, p. 315): 'Thus climate change has predominantly been framed as an environmental problem, rather than as a problem concerning people's needs, rights, and values.'

Although it may not be clear in popular discourse, vulnerability to climate change impacts (events and disasters) are socially determined (Pelling and Dill 2009; Wisner et al. 2012; Di Baldassare et al. 2018). Disasters can impact political institutions and actors, but the organisation and functioning of political institutions can influence the form, extent and distribution of effects of natural hazards, climate related events and disasters. Sowers et al. (2011, p. 600) have stated that climate change is a threat multiplier as its impacts will not be equally distributed, and that it is:

> … increasingly clear that climate change will interact with other social, economic and political variables to exacerbate social and political vulnerabilities.

This social construction of climate related events and disasters rejects the view that 'natural disasters' are external events imposed on society, with outcomes disconnected from social factors (Kelman 2020). From this social perspective there is a clear need to instead focus on connections between social factors, natural hazards, vulnerability and disasters. As a consequence, the case for identifying and look closer at broader governance issues and political aspects of climate change adaptation has been established through scholarship and research that have engaged with political values, concepts, institutions and practices.

Adaptation responses in policy and planning, however, reflect and produce social values and institutional influences as they concern decisions on allocating resources and distributing costs in society (and within and between natural systems) (Carlson and McCormick 2015; Sovacool et al. 2015; Tangney 2017). Understanding these influences is a motivation for identifying political factors in development issues, particularly negative effects for developing nations, regions and poor communities.

Other studies have also viewed vulnerability to climate change as being clearly socially determined (Di Baldassare et al. 2018). One expression of this is the problematisation by disaster scholars of the 'natural disaster' concept (cf., Kelman 2020). This critique centres on perceiving 'natural' disasters as disconnected from relevant social factors. Instead, scholars have stressed the need to focus on the connection between social factors, natural hazards and disasters (Wisner et al. 2012); i.e., the social construction of disasters. Other scholars have highlighted the political dimension of disasters as they can impact political institutions and actors but also emphasised how political institutions can determine vulnerabilities, as well as the direction and significance of change (Pelling and Dill 2009). Sowers et al. (2011, p. 600) have stated that the impacts of climate change will not be equally distributed, adding that climate change is a threat multiplier and that it is:

> ... increasingly clear that climate change will interact with other social, economic and political variables to exacerbate social and political vulnerabilities.

Baja and Granberg (2018) highlight the relevance of equity. As a consequence, the case for identifying the political aspects of climate change adaptation has been established through scholarship and research into natural hazards, risk management and adaptation with an emerging interest and engagement with power, political values, concepts, institutions and practices.

Adaptation has been taken as a technical challenge with a view, largely, of maintaining the status quo as much as practicable. Considering adaptation in largely technocratic terms achieved through scientific modes of operating runs the risk of undermining adaptation and producing negative social and environmental outcomes (i.e., maladaptation). As a result, the scholarship of adaptation research and policy is developing into a more vibrant and diverse field of research broadened through the increased recognition, inclusion and consideration of political values. Nonetheless, there remains considerable interest and tacit endorsement of adaptation planning and policy formulation, as expressed by agencies and researchers, as an essentially apolitical task and activity. Remling (2018), in a study of EU adaptation policy, makes an important contribution by examining the use of de-politicisation (2018, p. 490):

> Framing adaptation as an environmental (rather than a potentially controversial social or political) problem invites technocratic and managerial solutions ... renders social issues invisible in the adaptation discourse. It prevents the emergence of alternative visions and the solutions to achieve them. Consequently, climate change and adaptation become non-political.

A number of specific critiques of adaptation policy, policy-making and practical outcomes as technocratic have emerged from adaptation studies. Thus, the prevailing technocratic culture is being subjected to critical scrutiny, revealing its political (and economic) interests and making the case for a more inclusive, participatory decision making processes that consider the socio-political implications and outcomes of adaptation measures.

## Neoliberalism and Adaptation

Neoliberalism is the political ideology and program of economic liberalism; it seeks to reverse the growth of state intervention in economic life that characterised the post-Great Depression Keynesian (and Fordist) models for economic (and social) development and underpinned the ethos of welfare states (Eckersley 1993, 2004; Anderson and Leal 2001; Christoff 2005; Coffey and Marston 2013). Advocates identify, as virtues, markets' self-regulatory capacities, efficient economic allocation and use of resources, and equitable performance by rewarding superior enterprise and effort. Historically, neoliberalism arose in the 1970s and continues to be promoted and practiced around the world by governments, think tanks and international agencies. Manifestations of neoliberalism have included austerity programs, commodification of commons goods, corporatisation deregulation, free markets and privatisation. In some respects, neoliberalism began as an economic program combining market liberalisation and monetarist economic policy (such as balanced budgets and austerity). Where it is most strongly promoted, neoliberal economics is conjoined with socially conservative philosophies and goals. As an ideology it extends beyond classical economic theory, promoting the doctrine that markets are nearly always superior—both in an ethical sense and empirically in fulfilling social goals—to governments. Social benefit may be a rationalisation for baser motives however, as strong arguments are advanced that the greatest advocates of neoliberal practices have been those with vested interests, i.e., major corporations and their executives (Beder 2006). Indeed, some have argued that neoliberalism should be primarily regarded as a conservative political program and strategy (Harvey 2005).

Some critiques have found neoliberalism to be inadequately defined, often accompanied by observations that it cannot be located as a uniform practice (Harvey 2005; Castree 2010; Coffey and Marston 2013; Springer et al. 2016). Conservative commentators have offered that neoliberalism does not exist by claiming that there is no monolithic, universal and managed program, as such. Such critiques can be dismissed as a straw man, as nearly all critics of neoliberalism describe it as greatly influenced by local context and a variegated phenomenon (e.g., Harvey 2005). This debate, however, highlights the problem of recognising often intangible ideologies that are common and widespread. As Peck and Tickell comment on the 'new religion' of neoliberalism, promoting ideas such as free trade, flexible labour and individualism is now so common in politics '... that they hardly even warrant a comment in many quarters' (2002, p. 381). It is suggested that neoliberalism has developed into the dominant, even hegemonic, ideology of our times (Holborow 2015), i.e., the paradigm of contemporary capitalism (Buch-Hansen 2018).

Neoliberalism primarily concerns the place, role and reach of the state in economic and social matters (Springer et al. 2016). In popular discourse, neoliberalism is sometimes incorrectly depicted as eliminating government powers to the greatest extent practicable, whereas in practice it is a complex and highly varied reconfiguration and reimagination of state powers and relationships. Commentators describe the rollback and rollout aspects of neoliberalism, with suggestions that the predominance of the rollback in the 1980s has been largely supplanted by rollout activities (Peck and Tickell 2002). 'Rollback' refers to the destruction (and discrediting) of welfarist and collectivist state institutions. 'Rollout' neoliberalism refers to constructing state powers, institutions, regulations and for neoliberal governance and reregulation, including creating new markets, especially for public goods. Transformation of public sector agencies under the moniker of 'new public management' in the 1980s and 1990s expressed both the rollback and rollout agendas in the quest to 'modernise' government.

As a phenomenon of contemporary governance, neoliberalism has found expression in society-Nature relations, in how environmental issues are perceived and in the functioning of environmental public policy as market environmentalism. There is now a considerable critique of neoliberalism in environmental commentaries, with much directed as market environmentalism. Ciplet and Roberts (2017) identify four neoliberal features of international environmental governance: (1) Libertarian justice ideals, (2) Marketisation, (3) Governance by disclosure and volunteerism and (4) Exclusive decision-making. A number of studies have identified these influences in the international climate change regime, especially those of marketisation and GHG mitigation (e.g., Newell and Paterson 2010).

Adaptation falls under the influence of neoliberalism; adaptation activity and scholarship are primarily involved with issues of governance and public policy and thereby are inextricably interlaced with the key interests of the neoliberal program. Each of the aforementioned neoliberal features described by Ciplet and Roberts (2017) can be found in adaptation responses at the international and national levels (as shown in subsequent chapters in this volume).

Few specific studies, however, have examined the relationship between neoliberalism and adaptation directly (Remling 2018). Many political studies of adaptation activities clearly involve neoliberal themes and concepts without evoking neoliberalism, per se. Rather than dealing with climate change adaptation, there is considerably more interest in linking neoliberalism to various natural hazards and disasters (Oliver-Smith 2015). Attention has been directed at particular components of neoliberal approaches. Firstly, there is the diminution of government services and weakening of the state's ambitions, leaving individuals and communities to undertake the necessary actions that might have otherwise been the purview of the state. Oftentimes, this takes the form of promotion by states of individual enterprise and community initiatives in the form of volunteerism. Secondly, are the state efforts to create new spaces in which non-state actors can operate. A prominent discourse in this regard is that of 'disaster capitalism' (Klein 2007), wherein post-disaster activities promote neoliberal policies that serve to enrich vested interests and widen existing socio-economic inequities. Particularly relevant to international disaster aid is that NGOs can serve to open 'spaces' for neoliberal interests, giving them an ambiguous character. Thirdly, there are those state efforts designed to protect the interests of neoliberalism itself; i.e., an institutional framework to enable and secure neoliberal practices.

Perhaps the most widely scrutinized effects of a single expression of neoliberalism are those of the privatizing, federal funding reductions and other initiatives of the US Federal Emergency Management Agency (FEMA) under the G.W. Bush Presidency and its subsequent impacts on the inadequate preparations and incompetent and corrupt responses to Hurricane Katrina on New Orleans (Quigley 2008). In fact, no other single issue in this discourse has attracted the extent of commentary and scholarship of neoliberalism and 'disastrous recoveries' than post-Katrina New Orleans. This critique of neoliberalism also extended to its influence after many other disasters, such as in Aceh after the 2004 tsunami (Phelps et al. 2011) and Haiti's 2010 earthquake (Schuller 2016). Comfort et al. (1999) describe the effects of austerity programs as a consequence of the 'structural adjustments' on international banks on increasing vulnerabilities of citizens and state agencies in Nicaragua's response to Hurricane Mitch in 1998. There is also growing interest in viewing migration and 'climate refugees' as an expression of neoliberalism (e.g., Faber and Schlegal 2017).

Additional critiques of neoliberalism include those of neoliberalism as a language and discourse (Holborow 2015). One such critique holds that neoliberal ideology provides a language of abstractions, concepts and perceptions that embodies political values and preferences, infiltrating all levels of social life from the individual to the state.

Of political significance is the framing of climate change risks as natural phenomena. Nickel and Eikenberry (2007, p. 534) observe:

> Perhaps because they appear to be "natural," the disastrous effects of natural events such as Hurricane Katrina invoke a discourse that the disastrous effects of neoliberalism thus far have not.

As such, neoliberalism provides a way of framing vulnerability and providing symbols and concepts that shape the ways that climate change, natural hazards and risks are perceived and understood. Consequently, this discourse can shape every aspect of adaptation practice, including setting goals, expressing priorities, creating and assessing options and the like.

## Indigenous Peoples

Indigenous peoples are receiving greater attention in the adaptation discourse, having been overlooked in the earlier period of climate change discourse (c.f., IPCC 1996, 2014; ILO 2019). Ongoing research has highlighted their particular vulnerabilities and also the role of TEK in maintaining ecologically sustainable livelihoods. Politically, two issues have been prominent; firstly, the multiple disadvantages borne by indigenous peoples worsened by climate change and secondly, the inadequacy of their involvement in decision making regarding their interests. Of particular concern is that indigenous peoples generally suffer from many disadvantages socially, economically and politically onto which the negative impacts of climate change will be added. Pre-existing disadvantages include marginalisation from social and economic resources, the risks of hunger and famine, lack of political representation, poor health status, degradation of ecosystem services and resources and the negative effects of globalisation. In terms of environmental justice, the disparity between their very low contribution to global GHG emissions and their high vulnerabilities to climate change impacts could not be starker.

Multiple factors contribute to the high vulnerability of indigenous peoples' exposure to climate change. Poverty is a major factor and poverty levels are particularly high in these communities. Further, many of the ecosystem goods and services vital for survival are commons, managed collectively for these purposes (see, Box 3.3). Natural resource commons have proved to be vulnerable to appropriation through privatisation and theft, often under the impetus of globalisation and neoliberalisation. Geographical factors also contribute; Indigenous peoples are usually reliant on agriculture and primary production. Many of these productive systems are highly 'climate-sensitive' and vulnerable to climate change. Furthermore, indigenous peoples occupy disproportionally more vulnerable regions to climate change: coastal regions, deserts, high altitude regions, polar regions, the tropics and small islands.

A key issue facing indigenous peoples is that of displacement, relocation or migration because their homelands can no longer support them. Such so-called 'climate refugees' are likely to experience increased disadvantages, loss of human security and increased vulnerabilities. These peoples experience losses over and above those of material deprivation of goods and services; relocation can result in the loss of those aspects of life that provide identity, culture, security and spiritual support.

It is important not to consider indigenous peoples and their traditional ways of living exclusively in terms of victimhood as this distorts and obscures the significance of their ecologically sustainable livelihoods. TEK has given indigenous peoples a deep understanding and appreciation of their environment s and enabling them to cope with climatic variability over long periods. Such knowledge forms a valuable resource for dealing with climate change in ways that are ecologically sustainable (cf., Trundle 2018). Climate change, however, presents new challenges—especially where this involves disasters—where climatic conditions are without precedent.

There has been a tendency in policy to perceive indigenous peoples' knowledge and relationships with the natural environment as equivalents to those of modern industrial societies; i.e., in economic, rationalistic and materialistic ways. Such perceptions fail to appreciate indigenous peoples' emotional, cultural, spiritual and symbolic relationships with the environment and the central, integral and essential roles these play in lived experience (Trundle 2018). Approaches to adaptation politics focusing exclusively on material equity and representation in decision making overlook issues highly pertinent to the special needs of indigenous peoples. As Adger et al. (2011, p. 2) put it:

> We argue that climate change policy underemphasizes, or more often ignores completely, the symbolic and psychological aspects of settlements, places, and risks to them.

Reflections on the implications of this lesson for adaptation politics more generally might consider whether these have failed to grapple with the more intangible social values threatened by climate change in the cultural, emotional and symbolic realms.

**Box 3.3 Commons**

Commons resources and services are those managed, maintained and used collectively through institutions of cooperation. These institutions are founded on a property rights system that controls access and usage of resources of social value. Commons may appear to be similar to public goods, but a key difference is that access to commons can be controlled, whereas this is usually impossible or impracticable for public goods. Exemplars of public goods (valued for either for consumption or production) are breathable air, public knowledge, language and national defence. Public goods are also 'non-rivalrous' in that their use by one party does not diminish another's use; in contrast, one party's consumption of commons' goods and services reduces that available to others. Ecosystem goods and services on which most of the world's poor depend for water, food and fibre production, cultural and spiritual values are usually commons. These have been managed collectively so as to ensure the continuing stream of benefits and to avoid excessive harvesting, use and exploitation. As shared resources, commons are differentiated from private property and that owned and controlled by the state. Because natural resource commons have been managed over long spans of human existence—indeed, commons are arguably the reason for human existence up to the modern era—they are of particular interest as models for long-term managing and protecting ecosystems; i.e., models for environmental sustainability. Commons concepts are now being applied to other resource types, such as infrastructure. As such, commons are central to the problem of adaptation, but also to potential solutions.

See, Ostrom (1990) and Wall (2014).

## Adaptation and the State

As the discussion above has shown, adaptation is a highly complex concept and a collective action problem without clearly agreed upon responsibilities (Harris 2013; Moloney et al. 2018b). This points towards the need for collaboration and coordination that facilitates concerted action. It is clear, however, that public actors ('the state') on different levels have a special role, and that they are also giving increasing attention to adaptation responses. It follows that adaptation policy and measures engage public actors, especially on the regional and local levels (Rosenzweig et al. 2010, 2011; Bulkeley and Castán Broto 2013; Moloney et al. 2018a).

Since climate change poses both direct and indirect threats to public welfare in mature welfare states, governments have formal responsibilities (and electoral imperatives) to become engaged with adaptation (Gough and Meadowcraft 2013). Direct, climate-related risks include increased heat and water stress, sea level rise and more dramatic temperature changes. Indirect, climate-related risks include spill-over effects from climate impacts elsewhere. An OECD (2014) study of the potential of a major flood in Paris found that the infrastructure sector would account for 30–55% of the direct damages, but that 35–85% of the losses to businesses were caused by transport and electricity service disruptions (i.e., the indirect impacts of the flood). Other issues that can be perceived as indirect risk include migration as a result of water or food shortages and sometimes armed conflict leading to the breakdown of the global trade regime and significantly higher food prices (Mobjörk et al. 2016). These threats will put considerable pressure on welfare states, as Gough and Meadowcraft put it (2013, p. 493):

> These spillovers point to an inherent tension within welfare states, that in delivering entitlements to citizens they discriminate against non-citizens ... and can become 'fortress' welfare systems. Climate change, an ineluctably international phenomenon, will test the ability of national welfare states to internationalize and recognize collective responsibility elsewhere in the world.

Addressing certain climate change impacts requires the unique powers, 'reach' and legitimacy of state authority (Granberg and Glover 2014). It can require state powers for ensuring or enabling wider participation in planning and decision making, especially of community and advocacy groups. States also possess the resources and capacities for initiating and delivering adaptation programmes; they may also take up a role of ensuring more equitable distribution of resources and support where these might be highly unequal (such as between rural and urban locations, or cities with widely differing circumstances). There are also issues of scale that states can overcome. Local communities' adaptation responses may be inadequate in the face of larger-scale issues, spill-over effects across different jurisdictions and authorities, and shortfalls in matching resources and adaptation capacities to the scale of impacts or risks. Resolving such problems necessitates applying the authority, powers and roles of states (and inter-governmental coordination). Meeting adaptation obligations, however, can put pressure on state resources; funding and supporting adaptation may be to the detriment of competing core state functions (Biermann and Dingwerth 2004).

Taking a historical perspective, with the rise of the modern state and scientific knowledge in the contemporary era, adapting to climate changed significantly. *Firstly*, state agencies—through public policy instruments—assumed a role in organizing aspects of these responses at a societal scale (Massey and Huitema 2016). *Secondly*, a forward-looking and precautionary outlook prompted planning initiatives, especially for managing the risks of climate-related hazards (Grieving and Fleischhauer 2012). *Thirdly*, science provided insights into the future, and a range of technologies and tools that equipped public authorities to devise plans and undertake practices as part of the planning and risk management response (e.g., IPCC 2014). Therefore, measuring, mapping and describing climate-related hazards with potential social implications is within the interest of the state. Adaptation, therefore, became an additional realm within this established and broad institutional and instrumental setting. While adaptation is undertaken by households, communities, corporations, government entities, NGOs and others according to their specific needs, these activities invariably engage with states and their agencies. To this list should be added that an international dimension has emerged in adaptation, an activity that by definition engages states at the national and supra-national scale. This ranges from international negotiations involving (notably under the UNFCCC process), aid/financial support for international adaptation to programmes and projects undertaken with other nations. For some nations, notably those that are lesser developed and/or of smaller populations, only states have the institutional capacities to undertake larger-scale adaptation programmes (and organise additional support from external sources).

National and local governments and international agencies are giving adaptation increasing attention in public policy, planning and practice (IPCC 2001, 2007, 2012, 2014; Pelling 2011; World Bank 2011; EU 2013). This development is shown by an increasing number, and a widening scope, of public policies addressing climate change adaptation in the wake of recent extreme climate events and climate-related natural hazards (IPCC 2014).

In this process, adaptation policy and practice have had to move from narrow technical problem identification to broader policy interventions and this entails acknowledging and including political aspects and influences.

Part 2 of the volume deals with adaptation and politics in three chapters. Chapter 4 identifies two over-arching themes in adaptation politics, those of governance and sustainability. Chapter 5 examines prominent themes in adaptation politics; namely, the thematic areas that attracts most scholarly attention. Chapter 6 examines four prominent political debates in adaptation politics, each of which is informed by a specific theoretical and conceptual foundation; our categorisation comprises: institutional reform, political economy, environmental justice and political ecology. A concluding chapter, Chapter 7, that includes discussion, closes the volume.

## References

Adger, W. N. (2003). Social capital, collective action, and adaptation to climate change. *Economic Geography, 79*(4), 387–400.

Adger, W. N., Arnell, N. W., & Tompkins, E. L. (2005). Successful adaptation to climate change across scales. *Global Environmental Change, 15,* 77–86.

Adger, W. N., Barnett, J., Chapin, F. S., & Ellemor, H. (2011). This must be the place: Underrepresentation of identity and meaning in climate change decision-making. *Global Environmental Politics, 11*(2), 1–25.

Adger, W. N., Barnett, J., Brown, K., Marshall, N., & O'Brien, K. (2013). Cultural dimensions of climate change impacts and adaptation. *Nature Climate Change, 3,* 112–117.

Adger, W. N., Dessai, S., Goulden, M., Hulme, M., Lorenzoni, I., Nelson, D., et al. (2009). Are there social limits to adaptation. *Climatic Change, 93*, 335–354.

Agrawal, A., & Perrin, N. (2008). *Climate adaptation, local institutions, and rural livelihoods* (IFRI Working paper W08I-6). University of Michigan.

Agrawal, A., Perrin, N., Chhatre, A., Benson, C. S., & Kononen, M. (2012). Climate policy processes, local institutions, and adaptation actions: Mechanisms of translation and influence. *WIREs Climate Change, 3,* 565–579.

Anderson, T. L., & Leal, D. R. (2001). *Free market environmentalism.* New York: Palgrave.

Baja, K., & Granberg, M. (2018). From engagement to empowerment: Climate change and resilience in the city of Baltimore. In S. Moloney, H. Fünfgeld, & M. Granberg (Eds.), *Local action on climate change: Opportunities and constraints* (pp. 126–145). London and New York: Routledge.

Barnett, J. (2010). Adapting to climate change: Three key challenges for research and policy—An editorial essay. *WIREs Climate Change, 1* (May/June), 314–317.

Beder, S. (2006). *Free market missionaries.* London: Earthscan.

Berkes, F. (1999). *Sacred ecology: Traditional ecological knowledge and management systems.* Philadelphia and London: Taylor & Francis.

Biermann, F., & Dingwerth, K. (2004). Global environmental change and the nation state. *Global Environmental Politics, 4*(1), 1–22.

Birkmann, J., & von Teichman, K. (2010). Integrating disaster risk reduction and climate change adaptation: Key challenges—scales, knowledge, and norms. *Sustainability Science, 5*(2), 171–184.

Bosomworth, K., & Gaillard, E. (2019): Engaging with uncertainty and ambiguity through participatory 'Adaptive Pathways' approaches: Scoping the literature. *Environmental Research Letters, 14* (9), https://doi.org/10.1088/1748-9326/ab3095.

Buch-Hansen, H. (2018). The prerequisites for a degrowth paradigm shift: Insights from critical political economy. *Ecological Economics, 146,* 157–163.

Bulkeley, H., & Castán Broto, V. (2013). Government by experiment? Global cities and the governing of climate change. *Transactions of the British Institute of Geographers, 38*(3), 361–375.

Carlson, K., & McCormick, S. (2015). American adaptation: Social factors affecting new developments to address climate change. *Global Environmental Change, 35,* 360–367.

Castree, N. (2010). Neoliberalism and the biophysical environment. *Environment and Society: Advances in Research, 1,* 5–45.

Christoff, P. (2005). Out of chaos, a shining star? Toward a typology of green state. In J. Barry & R. Eckersley (Eds.), *The state and the global ecological crisis* (pp. 25–52). Cambridge: MIT Press.

Chu, E., Anguelovski, I., & Carmin, J. (2016). Inclusive approaches to urban climate adaptation planning and implementation in the Global South. *Climate Policy, 16*(3), 372–392. https://doi.org/10.1080/14693062.2015.1019822.

Ciplet, D., & Roberts, J. T. (2017). Climate change and the transition to neoliberal environmental governnce. *Global Environmental Change, 46,* 148–156.

Coffey, B., & Marston, G. (2013). How neoliberalism and ecological modernization shaped environmental policy in Australia. *Journal of Environmental Policy & Planning, 15*(2), 179–199. https://doi.org/10.1080/1523908X.2012.746868.

Comfort, L., Wisner, B., Cutter, S., Pulwarty, R., Hewitt, K., Oliver-Smith, A., et al. (1999). Reframing disaster policy: The global evolution of vulnerable communities. *Environmental Hazards, 1,* 39–44.

Dekens, J. (2007). *Local knowledge for disaster preparedness: A literature review.* Kathmandu: International Centre for Integrated Mountain Development (ICIMOD).

del Mar Delgado-Serrano, M., Oteros-Rozas, E., Ruiz-Mallén, I., Calvo-Boyero, D., Ortiz-Guerrero, C. E., Escalante-Semerena, R. I., et al. (2018). Influence of community-based natural resource management strategies in the resilience of social-ecological systems. *Regional Environmental Change, 18*(2), 581–592.

Dewulf, A. (2013). Contrasting frames in policy debates on climate change adaptation. *WIREs Climate Change, 4,* 321–330.

Di Baldassare, G., Nohrstedt, D., Mård, J., Burchardt, S., Albin, C., Bondesson, S., et al. (2018). An integrative research framework to unravel the interplay of natural hazards and vulnerabilities. *Earth's Future, 6,* 305–310.

Eckersley, R. (1993). Free market environmentalism: Friend or foe? *Environmental Politics, 2*(1), 1–19.

Eckersley, R. (2004). *The green state: Rethinking democracy and sovereignty.* Cambridge: MIT Press.

Eriksen, S. E., Nightingale, A. J., & Eakin, H. (2015). Reframing adaptation: The political nature of climate change adaptation. *Global Environmental Change, 35*, 523–533.

EU. (2013). *The EU strategy on adaptation to climate change*. Brussels: The European Union.

Faber, D., & Schlegal, C. (2017). Give me shelter from the storm: Framing the climate refugee crisis in the context of neoliberal capitalism. *Capitalism Nature Socialism, 28*(3), 1–17.

Fabricius, C., & Koch, E. (Eds.). (2004). *Rights, resources and rural development: Community-based natural resource management in Southern Africa*. London: Earthscan.

Farber, D. A. (2013). Issues of scale in climate governance. In J. S. Dryzek, R. B. Norgaard, & D. Schlosberg (Eds.), *The Oxford handbook of climate change and society* (Paperback ed., pp. 479–503). Oxford: Oxford University Press.

Fraser, N. (2001). Recognition without ethics. *Theory, Culture & Society, 18*(2–3), 21–42.

Füssel, H.-M., & Klein, R. J. T. (2006). Climate change vulnerability assessments: An evolution of conceptual thinking. *Climatic Change, 75*(3), 301–329.

Gifford, R. (2011). The dragons of inaction: Psychological barriers that limit climate change mitigation and adaptation. *American Psychologist, 66*(4), 290–302.

Gough, I., & Meadowcraft, J. (2013). Decarbonizing the welfare state. In J. S. Dryzek, R. B. Norgaard, & D. Schlosberg (Eds.), *The Oxford handbook of climate change and society* (Paperback ed., pp. 490–503). Oxford: Oxford University Press.

Granberg, M., & Glover, L. (2014). Adaptation and maladaptation in Australian national climate change policy. *Journal of Environmental Policy & Planning, 16*(2), 147–159.

Granberg, M., & Nyberg, L. (2018). Climate change adaptation, city competitiveness and urban planning in the city of Karlstad, Sweden. In S. Moloney, H. Fünfgeld, & M. Granberg (Eds.), *Local action on climate change: Opportunities and constraints* (pp. 111–125). London and New York: Routledge.

Granberg, M., Bosomworth, K., Moloney, S., Kristianssen, A-C., & Fünfgeld, H. (2019). Can regional-scale governance and planning support transformative adaptation? A study of two places. *Sustainability, 11*(24), 6978. https://doi.org/10.3390/su11246978.

Grieving, S., & Fleischhauer, M. (2012). National climate change adaptation strategies of European states from a spatial planning and development perspective. *Journal of European Planning Studies, 20*(1), 27–48.

Grothmann, T., & Patt, A. (2005). Adaptive capacity and human cognition: The process of individual adaptation to climate change. *Global Environmental Change, 15*(3), 199–213.

Harris, P. G. (2013). *What's wrong with climate politics and how to fix it*. Cambridge: Polity Press.

Harvey, D. (2005). *A brief history of neoliberalism*. Oxford: Oxford University Press.

Head, B. (2014). Evidence, uncertainty, and wicked problems in climate change decision making in Australia. *Environment and Planning C: Government and Policy, 32,* 663–679.

Hobsbawm, E. (1996). Identity politics and the left. *New Left Review, 217,* 38–47.

Holborow, M. (2015). *Language and neoliberalism*. London: Routledge.

Houde, N. (2007). The six faces of traditional ecological knowledge: Challenges and opportunities for Canadian co-management arrangements. *Ecology and Society, 12*(2), 34.

ILO. (2019). *Indigenous peoples and climate change: Emerging research on traditional knowledge and livelihoods*. Geneva: ILO.

IPCC (Intergovernmental Panel on Climate Change). (1996). *Climate change 1995: Impacts, adaptations, and mitigation of climate change: Scientific-technical analyses*. Cambridge: Cambridge University Press.

IPCC. (2001). *Climate change 2001: Impacts, adaptation and vulnerability*. Cambridge: Cambridge University Press.

IPCC. (2007). *Climate change 2007: Impacts, adaptation and vulnerability*. Cambridge: Cambridge University Press.

IPCC. (2012). *Managing the risks of extreme events and disasters to advance climate change adaptation*. Cambridge: Cambridge University Press.

IPCC. (2014). *Climate change 2014: Impacts, adaptation and vulnerability*. Cambridge: Cambridge University Press.

Kelman, I. (2020). *Disaster by choice: How our actions turn natural hazards into catastrophes*. Oxford: Oxford University Press.

Keulertz, M., Sowers, J., Woertz, E., & Mohtar, R. (2018). The water-energy-food nexus in arid regions: The politics of problemsheds. In K. Conca & E. Weinthal (Eds.), *The Oxford handbook of water politics and policy* (pp. 167–196). Oxford: Oxford University Press.

Klein, N. (2007). *The shock doctrine: The rise of disaster capitalism*. New York: Metropolitan Books.

Leonard, S., Parsons, M., Olawsky, K., & Kofod, F. (2013). The role of culture and traditional knowledge in climate change adaptation: Insights from East Kimberley, Australia. *Global Environmental Change, 23,* 623–632.

Makondo, C. C., & Thomas, D. S. G. (2018). Climate change adaptation: Linking indigenous knowledge with western science for effective adaptation. *Environmental Science & Policy, 88,* 83–91.

Manuel-Navarrete D., Pelling, M., & Redclift, M. (2009). *Coping, governance, and development: The climate change adaptation triad* (Environment, Politics and Development Working Paper Series, WP18). Department of Geography, King's College London.

Marshall, G. R. (2015). Polycentricity, subsidiarity and adaptive efficiency. In *Ostrom workshop in political theory and policy analysis*. Bloomington: Indiana University.

Massey, E., & Huitema, D. (2016). The emergence of climate change adaptation as a new field of public policy in Europe. *Regional Environmental Change, 16,* 553–564.

McEvoy, D., & Fünfgeld, H. (2013). Resilience and climate change adaptation: The importance of framing. *Planning Practice & Research, 28*(3), 280–293.

Measham, T. G., Preston, B. L., Smith, T. F., Brooke, C., Gorddard, R., Withycombe, G., et al. (2011). Adapting to climate change through local municipal planning: Barriers and challenges. *Mitigation and Adaptation of Strategies for Global Change, 16*(8), 889–909.

Mobjörk, M., Gustafsson, M.-T., Sonnsjö, H., van Baalen, S., Dellmuth, L. M., & Bremberg, N. (2016). *Climate-related security risks: Towards an integrated approach.* Solna: Stockholm International Peace Research Institute (SIPRI).

Moloney, S., & Fünfgeld, H. (2015). Emergent processes of adaptive capacity building: Local government climate change alliances and networks in Melbourne. *Urban Climate, 14*(1), 30–40.

Moloney, S., Fünfgeld, H., & Granberg, M. (Eds.). (2018a). *Local action on climate change: Opportunities and constraints.* London and New York: Routledge.

Moloney, S., Fünfgeld, H., & Granberg, M. (2018b). Climate change responses the global to local scale: An overview. In S. Moloney, H. Fünfgeld, & M. Granberg (Eds.), *Local action on climate change: Opportunities and constraints* (pp. 1–16). London and New York: Routledge.

Moya, P. M. L. (2000). Introduction: Reclaiming identity. In P. M. L. Moya & M. R. Hames-García (Eds.), *Reclaiming identity: Realist theory and the predicament of oostmodernism* (pp. 1–26). Berkeley, Los Angeles and London: University of California Press.

Nakashima, D. J., Galloway McLean, K., Thulstrup, H. D., Ramos Castillo, A., & Rubis, J. T. (2012). *Weathering uncertainty: Traditional knowledge for climate change assessment and adaptation.* Paris: UNESCO and Darwin, UNU.

Newell, P., & Paterson, M. (2010). *Climate capitalism: Global warming and the transformation of the global economy.* Cambridge: Cambridge University Press.

Nickel, P. M., & Eikenberry, A. M. (2007). Responding to "natural" disasters: The ethical implications of the voluntary state. *Administrative Theory & Praxis, 29*(4), 534–545.
Nightingale, A. J. (2017). Power and politics in climate change adaptation: Struggles over authority and recognition in the context of political instability. *Geoforum, 84,* 11–20.
Nyong, A., Adesina, F., & Osman Elasha, B. (2007). The value of indigenous knowledge in climate change mitigation and adaptation strategies in the African Sahel. *Mitigating and Adaptation Strategies for Global Change, 12*(50), 787–797.
O'Brien, K., Eriksen, S., Nygaard, L. P., & Schjolden, A. (2007). Why different interpretations of vulnerability matter in climate change discourses. *Climate Policy, 7*(1), 73–88.
O'Brien, K. L., & Wolf, J. (2010). A values-based approach to vulnerability and adaptation to climate change. *WIREs Climate Change, 1*(2), 232–242.
OECD. (2014). *Seine Basin, île-de-France, 2014: Resilience to major floods.* Paris: OECD.
Oliver-Smith, A. (2015). Conversations in catastrophe: Neoliberalism and the cultural construction of disaster risk. In F. Krüger, G. Bankoff, T. Cannon, B. Orlowski, & E. L. F. Schipper (Eds.), *Cultures and disasters: Understanding cultural framings in disaster risk reduction* (pp. 37–52). London and New York: Routledge.
Ostrom, E. (1990). *Governing the commons: The evolution of institutions for collective action.* Cambridge: Cambridge University Press.
Paterson, B., & Charles, A. (2019). Community-based responses to climate hazards: Typology and global analysis. *Climatic Change, 152,* 327–343. https://doi.org/10.1007/s10584-018-2345-5.
Peck, J., & Tickell, A. (2002). Neoliberalizing space. *Antipode, 34*(3), 380–404.
Pelling, M. (2011). *Adaptation to climate change: From resilience to transformation.* London: Routledge.
Pelling, M., & Dill, K. (2009). Disaster politics: Tipping points for change in the adaptation of socio-political regimes. *Progress in Human Geography, 34*(1), 21–37.
Phelps, N. A., Bunnell, T., & Miller, M. A. (2011). Post-disaster economic development in Aceh: Neoliberalization and other economic-geographic imaginaries. *Geoforum, 42,* 418–426.
Quigley, W. P. (2008). What Katrina revealed. *Harvard Law & Policy Review, 2,* 361–384.
Remling, E. (2018). Depoliticizing adaptation: A critical analysis of EU climate adaptation policy. *Environmental Politics, 27*(3), 477–497.
Rosenzweig, C., William, S., Stephen, A. S., & Mehotra, S. (2010). Cities lead the way in climate-change action. *Nature, 467*(7318), 909–911.

Rosenzweig, C., Solecki, W., Hammer, S., & Mehrotra, S. (2011). Urban climate change in context. In C. Rosenzweig, C. Solecki, W. Hammer, & S. Mehrotra (Eds.), *Climate change and cities: First assessment report of the Urban Climate Change Research Network* (pp. 3–11). Cambridge University Press.

Schuller, M. (2016). *Humanitarian aftershocks in Haiti*. New Brunswick: University of Rutgers Press.

Schulz, K., & Siriwardane, R. (2015). *Depoliticised and technocratic? Normativity and the politics of transformative adaptation* (Earth System Governance Working Paper No. 33). Lund and Amsterdam: Earth System Governance Project.

Sovacool, B. K., Linnér, B.-O., & Goosite, M. E. (2015). The political economy of climate change adaptation. *Natural Climate Change, 5*(July), 616–618.

Sovacool, B. K., & Brown, M. A. (2017). Scaling the policy response to climate change. *Policy and Society, 27*, 317–328.

Sowers, J., Vengosh, A., & Weinthal, E. (2011). Climate change, water resources, and the politics of adaptation in the Middle East and North Africa. *Climatic Change, 104*, 599–627.

Springer, S., Birch, K., & MacLeavy, J. (Eds.). (2016). *The handbook of neoliberalism*. London: Routledge.

Storbjörk, S. (2007). Governing climate adaptation in the local arena: Challenges of risk management and planning in Sweden. *Local Environment, 12*(5), 457–469.

Swyngedouw, E. (2010a). Apocalypse forever? Post-political populism and the spectre of climate change. *Theory, Culture and Society, 27*(2–3), 213–232.

Swyngedouw, E. (2010b). Whose environment? The end of nature, climate change and the process of post-politicization. *Ambiente & Sociedade, XIV*(2), 69–87.

Tangney, P. (2017). *Climate adaptation policy and evidence: Understanding the tensions between politics and expertise in public policy*. London: Routledge.

Tompkins, E. L., & Adger, W. N. (2004). Does adaptive management of natural resources enhance resilience to climate change? *Ecology and Society, 9*(2), 10. http://www.ecologyandsociety.org/vol9/iss2/art10/.

Trundle, A. (2018). Governance and agency beyond boundaries: Climate resilience in Port Vila's peri-urban settlements. In S. Moloney, H. Fünfgeld & M. Granberg (Eds.), *Local action on climate change: Opportunities and constraints* (pp. 35–52). London: Routledge.

Unwin, K., & Jordan, A. (2008). Does public policy support or undermine climate change adaptation? Exploring policy interplay across different scales of governance. *Global Environmental Change, 18*(1), 180–191.

Wall, D. (2014). *The commons in history: Culture, conflict and ecology*. Cambridge, MA and London, UK: MIT Press.

Wilson, E. (2007). Adapting to climate change at the local level: The spatial planning response. *Local Environment, 11*(6), 609–625.

Wisner, B., Gaillard, J. C., & Kelman, I. (2012). Framing disaster: Theories and stories seeking to understand hazards, vulnerability and risk. In B. Wisner, J. C. Gaillard, & I. Kelman (Eds.), *The Routledge handbook of hazards and disaster risk reduction* (pp. 18–34). London: Routledge.

World Bank. (2011). *Guide to climate change adaptation in cities*. Washington, DC: The World Bank.

PART II

# Adaptation as Politics

CHAPTER 4

# Adaptation Politics in Context: Governance and Sustainability

**Abstract** This chapter examines two underlying contextual issues of adaptation politics: Governance and sustainability. Here, the focus is on governance, not just governments. A significant change in context is the change in perception that adaptation is not only a local issue, but now is multi-scalar. Public policy is a central aspect of governance, but the boundary between public and private adaptation is contested, especially regarding funding and support. Insurance is being promoted as an important adaptive measure and this is discussed. Sustainability deals with three topics that underpin environmental politics discourse: utilitarian versus spiritual evaluations, technological change versus behavioral change and rejections of economic growth. These topics provide the subject matter for the prominent political debates in adaptation politics, each of which is informed by a specific theoretical and conceptual foundations.

**Keywords** Governance · Institutions · Market failures · Public policy · Politics · Technocracy · Sustainability

As described in Part I of this volume, adaptation is a dynamic field driven by exogenous and endogenous factors. Adaptation does not take place in isolation from other policy items and decisions and it is recognized that adaptation policy needs to be understood in relation to the broader policy



L. Glover and M. Granberg,
*The Politics of Adapting to Climate Change*,
https://doi.org/10.1007/978-3-030-46205-5_4

context of other existing policies and political priorities (Adger et al. 2005; Unwin and Jordan 2008; Granberg and Glover 2014; Moloney and Fünfgeld 2015; Granberg and Nyberg 2018; Moloney et al. 2018a, b). Understanding the framing of issues and the relationships and competition between policies and policy issues are important when analysing policy processes (Mettler and SoRelle 2014; Bacchi and Goodwin 2016), as these elements influence the scope of, and what can be included in, a specific adaptation policy (Granberg and Glover 2014; Granberg and Nyberg 2018).

To begin the second part of this volume on adaptation as politics in context, this chapter examines two underlying contextual issues, namely those of adaptation governance and sustainability. As is often the way with such background and contextualising material, it does not usually make an explicit appearance in the adaptation discourse. Several reasons might explain this absence, such as for experts, these matters are already known and for pragmatic purposes, such as in policies, plans and programmes, it is deemed unnecessary. Understanding the root cause of adaptation politics is, however, aided by knowing something of the assumptions, concepts, debates, ideas and value differences underpinning these politics. In other words, these contextual materials give an insight into the deeper political discourse that shapes the surficial public discourse.

## Political Context I: Governance

'Governance' refers here to governing adaptation activities and to the characteristics of the system or theory underlying those activities; namely, it is the outcome of governing by actors in the social, political and administrative spheres (Bulkeley and Betsill 2003; Eckersley 2004; Renn 2008; Bauer et al. 2012; Moloney et al. 2018b; Patterson 2018). As the collective societal effort to undertake adaptation, governance extends well beyond the purview of governments and involves specialist institutions, rules, organisations and engagements with public and private actors. Advocates of 'good governance' associate it with desirable features, usually evoking governance by democratic governments, including the rule of law, accountability of decision-makers, participation in decision making, transparency, responsiveness, consensus and equitable outcomes (Nanda 2006). While it is reasonable for those living in liberal democratic nations to make such assumptions, it is worth bearing in mind that classical political science identifies a number of competing forms of governance to democratic rule, including monarchies, dictatorships, oligarchies, totalitarian regimes and authoritarianism. To such lists can be added community-based/collective approaches to governance, of which there are many forms (particularly for managing commons, see, e.g., the Common-Pool-Resource database, https://dlc.dlib.indiana.edu/dlc/contentguidelines).

A notable change in the thinking and policy formulation in adaptation concerns scale and governance (Farber 2013; Sovacool and Brown 2017; Moloney et al. 2018a; Granberg et al. 2019). In the early days of climate change policy, the accepted view was that while GHG mitigation was broad-scale issue of international relations, adaptation was a local matter, where needs and responses were dictated by local environmental factors, geography, culture, institutions and the like (i.e., 'place-based'). Since climate change impacts can occur at all scales, from the global to the local and down to individual households and enterprises, the realisation has emerged that adaptation responses should match the scale at which problems can occur (Marshall 2015). Different forms of governance, policy processes and community engagement are involved for different scale problems, such as strategies ranging from urban planning, to large-scale infrastructure provision to technology development. Accordingly, policy and scholarly attention has had to expand and consider governance at larger scales and include the challenges of multi-scale governance. There are a host of climate change impacts extending across political boundaries making manifest the limitations of responses bounded within single governance institutions or levels. Examples of cross-boundary issues created or worsened by climate change include:

- Cross-border pollutant or weather event issues physically separate cause and effects occur in different jurisdictions
- Resource harvesting, such as the impacts of water storage or extraction that imposes costs on downstream catchments
- Population movements or migration, and
- Political instability in neighbouring territories prompted by disruptions to natural resource harvesting systems/agriculture.

Enlarging the scope of governance concerns has altered the context of adaptation governance. Whereas previously, public policies for adaptation were assigned to existing governmental jurisdictions, wherein the task of national and state/provincial governments was largely taken up with providing the means for activities by local jurisdictions (including financing, technical and advisory services, and information provision). Adaptation governance now encompasses society at multiple scales and, importantly, inter-scale governance activities (Marshall 2015; Sovacool and Brown 2017; Moloney et al. 2018b).

### *Governance: Public Policy*

Government is prominent in adaptation for several reasons, some of which are unequivocal and others, contested. As Fieldman (2011, p. 166) explains, a prominent rationale, perhaps also the most obvious, is that the role of states to financially support adaptation is 'crucial', since other funding sources 'show no promise of adequacy'. Evidence supports this claim, e.g., Patterson's (2018) survey of urban water sector innovations in adaptation measures in 96 global cities showed relatively little activity by market actors in the private sector. Public policy is, in this regard, an essential instrument in mobilising financial support for adaptation.

Much of this policy activity is organised around protection of social welfare from climate related disasters and other climate impacts (i.e., socio-economic equity arguments), providing information/advice to enable adaptive responses and protecting public goods (Fankhauser et al. 1999; Tompkins et al. 2010). Public policy is also deployed to those functions that markets, expressing either corporate and civil society interests, are inappropriate, unable or unwilling to provide. These public policy interventions are directed at overcoming market failures in adaptation, including environmental externalities, missing/incomplete markets (including lack of sufficient information), monopoly markets, inefficient markets, 'de-merit' goods and perverse incentives (including maladaptive responses) (Adger 2003; Barnett and O'Neill 2013; Fieldman 2011; Granberg and Glover 2014; Mol 2016). Barriers in the cultural or economic realms encompass a wide array of issues, many of which require the breadth and depth of state powers and resources to overcome through policy initiatives and their implementation. Public authorities also have legislative and/or regulatory responsibilities for managing ecosystems and services whose natural adaptive capacity or resilience to change is insufficient for avoiding unacceptable loss es or damages.

While the case for public policy governance is evident in regard to the roles that can be considered the largely exclusive role of states, the boundaries of these roles are questioned. As DRR experiences have shown, there are a number of services that can be provided either, or by both, the public and private sectors. Further, public sector governance need not be provided by the public sector; in the neoliberal era, there has been extensive use of corporations to provide such services under state direction. Governance can also occur through economic institutions expressed through markets. Such choices may be resolved using economic analysis to identify the most cost-effectives means of service delivery, but governance clearly involves a far wider array of considerations than the merely economic. Determining the respective roles for states, markets and individuals is a difficult and complex issue.

It is clear that adaptation, overall, will not be inexpensive (at least not in the short run) and this highlights the need to address the issue of how financial responsibility for funding adaptation should be allocated (Farber 2013). Governments are already bearing considerable additional investment costs in infrastructure, such as such as for drainage, flood protection and coastal protection around the world (Moloney et al. 2018a). Resources, funding and support, however, are critical issues and ones that can only become more prominent in the adaptation discourse in the future and posing increasing pragmatic and political problems for governments. As Pelling (2011, pp. 12–13) states:

> With the cost of climate change increasing and adaptation being increasingly demanded, meeting the funding gap for adaptation in the short term is a key challenge.

Hence, the question how this gap should be handled and 'Who should pay?' is valid. One answer floated in the literature is the 'beneficiary pays' principle that simply requires that individuals and organizations benefitting from adaptation measures have the responsibility to pay for them (Page 2012; Farber 2013). In line with this argument, the view is that if you chose to live or conduct business in an area with more intense climate risks you should be responsible for the costs induced and not burden your fellow citizens, businesses or the state with these costs. At first glance this can seem as a plausible and functional answer to the issues of financing and responsibility, but critics argue that this principle is both implausible and unjustifiable (Huseby 2015; Barry and Kirby 2017). Notwithstanding the conditions where states have assumed responsibility for adaptation (outlined above), there are many aspects where the matter of risk taking and responsibilities for adaptation funding are open questions. This issue of who should bear the financial burden of adaptation is contested, with tensions between the interests of individuals, businesses and states.

## Governance: Markets and Insurance

An exception to the strong focus on public actors is the role of climate insurance, even though insurance also be can provided by public entities (Surminski and Eldridge 2017). It is evident that among private corporations, that insurance companies have the clearest financial incentives as they face direct exposure to the damage costs caused by climate change (Dryzek et al. 2013; Surminski et al. 2016). Insurance has also been central to the climate impact discourse of the UNFCCC and '... is now anchored in major policy arenas as one tool to address the risk of climate change, including the Paris Agreement and the Sendai Framework for Disaster Risk Reduction' (Schäfer et al. 2019, p. 318). As an adaptation response, widespread use of insurance (including for property, liability, health and life) offers a way to spread the risk and costs of climate change impacts (see, Art. 8 of the Paris Agreement). Insurance facilitates spreading the risk of losses to climate change impacts through time, geographically and between a mixture of private and corporate entities. In 2015, the G7 leaders launched the Initiative on Climate Risk Insurance that aimed to insure 400 million, currently uninsured, people in developing countries. In addition to compensation for losses through insurance claims, other benefits of insurance are investments by insurance firms in risk reduction and providing incentives for risk reduction (i.e., by 'risk pricing') (Granberg 2019). Insurance spreads the risks among insurance holders, facilitates collective management of losses and mitigates the impacts on individual actors in ways that limit the need for costly individual action (Schäfer et al. 2019).

As such, insurance has attracted considerable attention as an adaptation measure, but it has a number of specific political issues and implications. Attractions of insurance options are, however, offset by a number of serious deficiencies (Mills 2005; Duus-Otterström and Jagers 2011; Surminski et al. 2016; Schäfer et al. 2019):

- Efficacy concerns arise through the 'exceptional' impacts that climate change produces in natural systems. For insurers and the insured, climate change disrupts the construct of risk prediction that spreads risk across time and space and between insurance firms, governments and households
- Because insurance from insurance firms is relatively expensive, it is beyond the means of the poor; in the event of society-wide financial losses from climate change, insurance firms pay-outs to claimants have a regressive socio-economic effect
- Insuring the poor is highly challenging in the face of uncertain risks, low formal education levels and weak financial infrastructures/services in developing nations
- High uncertainty of extreme climate events and the possibility of high pay-outs are of rising concern to insurance firms (especially in view of previously inconceivable, but realised, disasters) and has the possibility to bankrupt some insurers
- Climate change risks may make insurance become unavailable (or greatly reduce coverage) in some locations (the 'uninsurable addresses'); in the case of property insurance, coverage loss depresses property values
- Insurers can use insurance terms to create pressure on public actors by refusing to insure properties in areas facing climate risks
- Insurance (especially if publicly subsidised) may be a disincentive to undertake effective risk reduction or take other more effective adaptive measures, and
- Insurance is not necessarily the optimal economic adaptation solution (being characterised by relatively high transaction costs).

For governments, widespread insurance as adaptation is itself an additional risk. In the event of large-scale catastrophic events, the viability of individual insurance firms will be threatened and their failure would constitute maladaptation on a large scale. Governments will have to face the economic and welfare implications of insurance firm failure through climate change events, a consequence that might prompt greater public policy interest and involvement in the insurance business.

## Political Context II: Sustainability

Traditional textbook depictions of politics are often of limited use in understanding the politics of sustainability and adaptation (Scoones 2016; Olsson 2018). Despite a multitude of views and opinions in the scholarly literature, politics is commonly presented as a spectrum of left-wing and right-wing views or political categories of economic and social beliefs arranged into matrices yielding quadrants along axes of radical/conservative and democratic/authoritarian continuums, identified as communist, fascist, liberal and so on.

Environmental politics has emerged as a field of inquiry that responds to, and draws upon, classical political philosophies—but in many instances, such traditional political ideologies are poorly equipped for interrogating Nature-society relations (Carter 2009; Dobson 2016; Gabrielson et al. 2019). Consequently, environmental politics has necessarily evolved to become a distinctive branch of political philosophy, theory and scholarship, with a particular set of issues, concerns, concepts and discourses (cf., Dryzek 1987; Dobson 1992; Eckersley 1992; Plumwood 1993; Olsson 2018; Blühdorn 2019). Further, environmental politics has spawned its own political movements, activists, organisations and political parties (cf., Richardson and Rootes 1995; Barry 1999; Carter 2009). Although sometimes disputed, it is argued that environmental politics has also developed distinctive ideologies (Eckersley 1992; Doherty 2005).

Underlying many of the different opinions over environmental issues in government deliberations, protest movements, public controversies and other realms of public discourse are the fundamental views that shape environmental politics. Oftentimes, these fundamental views are based on environmental values that give rise to the contrasting positions between key stakeholders and creating debates distinctive to environmental politics. These differences have been central to scholarly debate in environmental politics, such as featuring in controversies over the meaning and application of sustainable development (Adger and Jordan 2009; Carter 2009; Scoones et al. 2015; Sovacool and Linnér 2016). Throughout the history of environmentalism there have been a number of essential thematic debates over environmental values. Arguably, there are a multitude of such debates in environmentalism, but below three of the most essential related to sustainability are described: utilitarian vs. spiritual evaluations, technological change vs. behavioural change, and economic growth.

### *Sustainability: Utilitarian vs. Spiritual Evaluations*

Modern environmentalism (i.e., that of industrial civilisations) has its immediate roots in historical controversies in the latter Nineteenth and early Twentieth centuries with public debates over using and evaluating natural resources and related phenomena, particularly at the landscape scale, such as native forests, waterways, native fauna with economic value and areas of scenic beauty and cultural significance (O'Riordan 1976; McGormick 1989; Ponting 1992; Grove 1995; Pepper 1999). In the ensuing public debates, two distinct voices emerged, and these have remained salient in environmental disputes—and how we think about them—ever since: *conservationism* and *preservationism* (Carter 2009). These polar positions can be a shorthand representing a constellation of opposing views about the natural world (albeit that these terms have now fallen out of favour) and remain helpful guides to these differences wherein conservationism represents anthropocentrism and biocentrism and preservationism represents ecocentrism.

Although adopted in popular discourse to cover any form of environmental protection (as 'wise use') or even more loosely taken to merely mean an awareness of an environmental issue, here, we use the term and concept of conservation in its original, narrower and more precise meaning. Conservation is a valuation of the natural world based on its usefulness to society, i.e., its utility (Carter 2009). Political scholars trace this theme back to the classical proponents of utilitarianism, and its Nineteenth century proponents, such as Jeremy Bentham and John Stuart Mill, with the maxim of 'bringing the greatest good to the greatest number'. From this view, the concept of conservation emerged. In short, society should protect certain natural resources, ecosystems and ecosystem services because they are valuable to us. Under conservation, harvested forests should be replanted, fisheries should be harvested at sustainable rates, rivers should not be polluted past the point where they cannot accept future pollution and so on. As a goal and a practice, conservation aims to ensure a continued flow of benefits, or enable future benefits, to society from the natural world and employs the concepts such as sustainable use and sustainable yield.

Preservation, on the other hand, values the natural world on the basis of its intrinsic values, i.e., society has no right to diminish Nature on the basis of human interests (Carter 2009). Preservation's antonym is conservation; preservation is ecocentric, whereas conservation has a strong anthropocentric aspect (and 'biocentric', meaning that human and non-human life both have intrinsic value). Preservation is frequently tied to US writer Aldo Leopold and his 'land ethic' of the mid-Twentieth century (Lutz Warren 2006), but under ecocentrism and the Deep Ecology movement of latter decades, the span and scope of interest was greatly expanded. It is probably more accurate to recognise the wider range of sources for the cause of recognising values of Nature based on recognising spiritual, religious and other non-material values and that these arose from a wide array of cultures and sources around the world (see, Guha 2000). Such ecocentrism is also comprehensive in its concerns; whereas conservation usually focuses on selected species and specific locations, preservationist views embrace natural phenomena from the genetic to the ecosystem. As a rule, preservationists tend to place society 'outside' the environment; the social is typically taken as being artificial. A key theme is that the natural world is autonomous and self-organising. Albeit that the scholarship on this issue is dominated by the Western nations, many cultures have expressed spiritual regards of the natural world, and accordingly, spiritual values are not a monolithic concept.

Even though these are crude categories, they have some analytical power in explaining a key division in many environmental debates. Although conservation tends to focus on a smaller set of concerns, conservation problems are open to a wide range of solutions and are, in many senses, more amenable to negotiation and compromise. Much of the response to environmental issues has entailed differing views, approaches and techniques, and conservation tends to feature in approaches characterised by administrative rationalism, ecological modernisation, economic rationalism and sustainable development, attracting inputs from the fields of administration, economics, law and science. Preservation (or ecocentrism) has a much broader span of interests, but correspondingly entertains a far more limited set of solutions. Whereas conservation deals with practical compromises, preservation has far stronger spiritual, ethical and ethereal elements, and is associated with 'green consciousness', environmental philosophy and radical ecology.

Often sublimated in environmental debates, spiritual values can be the foundation of the most strongly held views about the environment (Taylor 2001; Hulme 2009). This debate between the utilitarian and the spiritual is still central to environmental politics, and in the context of legislation and regulation, policy formulation, planning and implementation, opposing views are often characterised as pragmatists and idealists. Preservation outcomes are relatively rare, but exemplars are biological reserves and prohibitions on some industrial compounds and practices. When 'preservationist' outcomes occur from environmental politics processes based in public policy, the reasoning is rarely based on spiritual values. In OECD nations, it is not unusual to find public agencies rationalising spiritual values using utilitarian reasoning and rare to have any species, ecosystems and locations protected because they possess intrinsic value. Industrial societies express instrumental and utilitarian valuations of the natural world as routine behaviour. In part, this may be due to shifting priorities and public debate away from issues, such as nature reserves and natural resource use and towards more urban concerns, such as ambient air pollution and climate change, but more fundamentally, is a marker of cultural values.

Allison (2015) described the spiritual values and symbolic importance of glaciers to local communities and the associated threats of climate change using examples from China, Nepal and Peru, noting that these threats have been largely ignored in the climate change discourse. Tschakert et al. (2019) produced a global reckoning of what they labelled as the 'intangible harms' of climate change, noting that most accounts of non-economic harms concern wealthy nations and have neglected those suffered by the poor. These authors note the difficultly in dealing with matters that are not—and perhaps should not—be quantified, but concern issues of identity, belonging in the contexts of place and culture.

### *Sustainability: Technological Change vs. Behavioural Change*

If utilitarian and spiritual valuations express two major rationales for environmental protection with greatly contrasting implications, this set of issues relates to how environmental protection can be achieved. Adaptation measures can be categorised in various ways, such as structural (e.g., technological, engineering and ecosystem-based), institutional (e.g., laws/regulations, policies and programmes, and economic instruments), and social (e.g., behaviour change, information provision and educational), see, e.g., Table 2.1. From a high-level perspective, the array of responses can be cleaved into broad categories. Dewulf (2013), amongst others, identified one framing of adaptation as a contrast between it as a 'tame technical problem' and 'wicked problem of governance'. A 'wicked' problem is basically a solution-resistant policy problem, while a 'tame" problem can be solved by policy measures/technology at hand (following the original Rittel and Webber (1973) concept). In relation to climate change, it has bees stated that the wicked (or super wicked) problem comprises four components: '… time is running out; those who cause the problem also seek to provide a solution; the central authority needed to address them is weak or non-existent; and irrational discounting occurs that pushes responses into the future' (Levin et al. 2012, p. 124). The concept of wicked problems has been problematised over the years (see Turnbull and Hoppe 2019) but can function as an illustration of the complexity of adaptation. Sovacool (2011), identified two principle paths to adaptation:

- *Hard* adaptive measures: are capital-intensive, large, complex and focusing on inflexible technology and infrastructure, and
- *Soft* adaptive measures: prioritize natural capital, community control, simplicity and appropriateness.

Sovacool based his thinking on Amory Lovins' and Herman Daly's perspectives on energy policy that stressed that the choice between the hard or the soft path had more to do with politics and policy than with technology, per se (Daly 1979; Lovins 1979). Lovins stated that '… each path entails a certain evolution of social values and perceptions …' (1979, p. 12). This debate harks back to the foundations of contemporary environmental thought. O'Riordan (1976) considered that environmental views could be 'technocentric' or 'ecocentric'. Indeed, two of contemporary environmentalism's 'founding fathers'—Paul Ehrlich (pro-behaviour change) and Barry Commoner (pro-technology change)—contested this very choice (see the discussion in, Feenberg 1999).

This debate can be applied to adaptation politics from two perspectives. *Firstly*, adaptation through applying technology. According to the IPCC (2014), engineering and technology options are the predominant adaptation response. Adaptation technologies are viewed as one of the main measures in preparing for and responding to climate change impacts, especially through technology transfer and diffusion in less developed nations (e.g., UNFCCC 2006; ADB 2014). Technology covers not only materials and equipment, but also includes technical and commercial information, and skills, training and knowledge for its use; 'adaptation technology' is a shorthand for what are often technological assemblages or systems. IPCC (2014) describes structural and physical adaptation options and groups technology with engineering and ecosystem-based adaptation. As the IPCC (2014, p. 846) explains, most engineering activities are '... expert driven, capital-intensive, large-scale, and highly complex'. These technologies are highly diverse, ranging from genetic techniques, crop and animal varieties, to food storage, to hazard mapping, early warning systems, heating, ventilation and cooling systems for buildings, as well as the more familiar engineered works, such as sea walls and levees, disaster shelters, water storages and irrigation systems. Most applications have been in the agriculture, coastal regions, disaster risk management, water resources and urban infrastructure sectors.

A focus on technological change and solutions is often labelled 'technocracy' (cf., Sartori 1987; Fischer 1990) or 'technocentrism' (cf., O'Riordan 1999; Bailey and Wilson 2009) and holds that science and engineering (through a technical elite) can resolve social and environmental problems using rational and instrumental logic. Technological solutions tend to have a greater interest in isolated problem solving but rarely deal with system-level environmental problems. Such critique is not a rejection of technological solutions, but of a singular reliance on them and the deterministic assumptions of technological optimism. Clearly, there are many opportunities for adaptation technologies, but optimism over their use needs to be tempered by recognising their existing barriers and limits. Typically, the limits to technology include its appropriateness to cultural settings, economic costs, effectiveness in different settings and trade-offs made in selecting technologies between competing choices involves.

In many cases, the limitations of technological adaptations express the strictures of path dependency. Large infrastructure soaks up available capital that might have been employed for other adaptation measures, with governments and corporations motivated to persist with its use in light of seeking returns on sunk capital. Further, such infrastructure often promulgates an associated assemblage of other technologies, investments, as well as political stakeholders and specialised institutions that, in combination, further lock in such technology choices.

Criticisms of specific technological adaptation projects, and of technocentric approaches, often constitute one or more of the following (and are increasingly identified as maladaptations). Firstly, there are direct social and environmental costs. Sometimes these are called 'unintended consequences' and are genuine surprises, but far more likely is that the technology proponents failed to value social and environmental phenomena at risk that could have reasonably foreseen through investigation. For example, Magnan et al.'s (2016) study of maladaptation refers to the case of the large-scale infrastructure investments in south–west coastal Bangladesh (a global 'hot spot' of extreme vulnerability to its inhabitants); these international agency and national investments have encouraged continued occupation (and perhaps encouraged inward migration), thereby exposing millions to the inevitable long-term hazards of climate and related impacts of the area. They describe these actions as "palliative adaptation", in that things are made better until they worsen.

Secondly, there are indirect environmental costs. These occur through technology's systemic effects, such as encouraging increased risk-taking behaviour, when more efficient/less expensive products and services encourage consumers to increase their net material consumption levels to the detriment of the environment (i.e., 'rebound effects') or when existing exploitative practices are 'locked-in' via technological developments. Barnett and O'Neil (2010) describe the construction of a desalination plant in Australia in response to long-term declines in water storage due to declining rainfall, but point out that the plant will add significantly to GHG releases, result in higher costs for drinking water to the detriment of the poor, is likely to erode self-sufficiency in household water conservation and has a high opportunity cost in the foregone options for wastewater treatment and increased domestic water storage. Further, the authors describe the path dependency implications of the plant that reduces the opportunities for alternative adaptive measures.

Thirdly, the indirect social costs. These take many forms and include contributing to socio-economic or political inequities by securing existing patterns of social and economic exploitation and inequality. Technology choices and their use in capitalist economies are largely dictated by market forces that preference investments offering the greatest economic return to investors, rather than those technology options with the greatest net social benefits. Technology markets are also geared to towards (private) ownership and private benefits arising from usage; investments for collective benefits usually depend on state support. Social inequality is increased as disadvantaged groups may not be able to take advantage of such technologies. More broadly, using economic metrics to determine adaptation funding priorities for technological adaptations can readily translate into funding support for large corporations and enterprises, choices that invariably reinforce existing economic relations and wealth distribution.

Large-scale public infrastructure investments for adaptation, often rationalised on the grounds of economic importance or protection of valuable assets, can be, in effect, transfers of public wealth to private interests—the full benefits of which may only accrue sometime in the future. Using public funds to build protective sea walls to protect valuable coastal real estate, for example, is effectively a subsidy to wealthy homeowners. Such investments also transfer or redistribute current or future risk via the agency of technology.

Risks of such outcomes may be ameliorated to varying degrees through appropriate technologies, community-based natural resource management and ecosystem-based approaches (sometimes called 'soft technologies'), although these often entail pre-modern technologies and approaches. *Secondly*, there is behavioural change (Gifford et al. 2011). Interventions through policy and related processes that can change individual or collective behaviour are very broad indeed, covering communication, education and information, as well those from institutions (notably planning, public policy, property rights, law and regulation) and from markets (including insurance, funding, microfinance, subsidies and taxes). Many of the behavioural measures can also be directed against the barriers to adaptation, changing behaviour that is impeded by existing policies, laws, institutions and the like. Behavioural change can be made possible through efforts understand the shaping of climate change perceptions, attitudes and valuations. Behavioural programs also run into limitations, including the inhibiting effects of entrenched values and (infra)structures, corporate interests, price and cost barriers, limits of available choices, demographic/geographic factors and subsidies to existing industries.

As with the utilitarian and spiritual perspectives, technocentrism and behavioural change represent contrasting worldviews that also express consistency around core values, priorities and normative goals. Technocentrism tends to highlight a commitment to economic growth and anthropocentric world views, utilising managerial perspectives that see environmental problems as resolvable without fundamental social change (or changes in social values) (see, Carter 2009). Much behaviouralist support is similarly reformist, both of the 'educational' and 'neoliberal' cast, but there is an ecologism variety that seeks more radical change in Nature-society relations and follows a transformative agenda (see, Dobson 1992). Despite such overlaps, the broader differences between technological and behavioural solutions to environmental problems are apparent and represent fundamental choices between different pathways.

### *Sustainability: Economic Growth*

Economic growth is a core concern of environmental sustainability—perhaps the primary concern—and is therefore a major theme in environmental politics (Blühdorn 2019). A seminal phase in environmental thinking was the rise of sustainability in the 1960s, 1970s and 1980s marked by the recognition of environmental issues as being constituent parts of wider ecological systems and processes, together with the realisation that social systems were subject to ecological laws of material throughput (WCED 1987). Prominent amongst these ecological laws was that natural non-renewable resources are limited and society is without substitutes for many vital ecosystem services, so that socio-economic systems are always ultimately subject to ecological constraints (WCED 1987; Arrow et al. 1995; Wackernagel et al. 2002). Daly (1990, p. 1) suggested that development as implying on-going quantitative economic growth within the finite global ecosystem 'be rejected as a bad oxymoron'. This reasoning produced 'the limits to growth' rationale that ran counter to conventional economic reasoning that not only considered economic expansion to be potentially limitless, but also used measures of economic growth (such as GDP) as the yardsticks of economic performance (Meadows et al. 2004).

Applying ecological principles, systems theory and other elements to economics gave rise to ecological economics following the scholarship of Nicholas Georgescu-Roegen (1971), Kenneth Boulding (1978), Herman Daly (1996) and others. A centrepiece of this activity was the concept of economic activity reconciled with ecological principles in the form of the 'steady-state economy' where economic growth is constrained, if not curtailed. Such an economy was clearly at odds with the conventional economist's endorsement of continual economic growth. A steady-state economy, therefore was one where renewable resource harvesting rates did not exceed those of regeneration, where non-renewable resource use rates do not exceed those at which substitute resources are discovered and that waste production does not exceed natural capacities of safe assimilation. This gave rise to idea that these resources and services constituted essential 'natural capital', the protection of which required their maintenance (i.e., the condition of sustainability). Such resource-based approaches to sustainability were grounded in ecological concepts, notably that of carrying-capacity; i.e., that a population's resource demands set the limits on population size, temporary 'overshoot' notwithstanding (Arrow et al. 1995).

With the rise of sustainable development, the place of economic growth was somewhat ambiguous. Largely, sustainable development began as the project to address environmental problems and the need for economic development to alleviate the global crisis of poverty and its attendant ills (such as promulgated by UNCED). Economic activity was to be restrained to avoid environmental damage. Some environmentalists were unconvinced, for as Daly and others pointed out, this model of development relied on greatly increasing resource consumption and waste generation and was therefore in violation of the ecological limits to growth (Arrow et al. 1995; Daly 1996, 2007; Meadows et al. 2004). An attempted reconciliation was the proposal that economic development could be secured through economic growth, provided that such growth was essentially qualitative improvements to economic productivity; i.e., that economic growth could be secured without greater material throughput. Whilst sound in theory—and despite the growth in sustainability resource usage, processing, products and services—the unassailable fact is that our era of globalisation has been one of greatly increasing resource use, mass consumption and waste output (as evidenced by various global and national sustainability indexes). Growth remains the indisputable characteristic of economic activity worldwide, prompting a more contemporary phenomenon of the 'de-growth' concept and associated social movements (see, Martínez-Alier et al. 2010).

Economic growth is politically an attractive way to address development needs for the obvious reason that the contentious issue of wealth distribution can be overlooked, whereby economic growth facilitates leaving the status quo of resource distribution within and between societies intact (Piketty 2014; Parr 2019). President John F. Kennedy's statement that 'a rising tide lifts all boats' is the aphorism *par excellence* that provides the normative social goal of economic growth (the common rejoinder is that you have to have a boat in order to be lifted). Conditions of economic growth in capitalist societies are nearly always proportionally more advantageous to the wealthy and the owners of capital, so that growth historically promotes inequality (Piketty 2014). If economic growth is to be stopped by the imposition of quantitative limits on consumption (such as would satisfy the requirements of de-growth), then the cause of addressing economic injustices in wealth distribution has to be addressed through wealth reallocation. In a sense, a steady-state economy reduces resource distribution in society to a zero-sum game, thereby evoking fundamental and difficult political challenges. Opposition to economic growth, a hallmark of strong forms of environmentalism, carries with it potent political implications and questions of the extent to which adaptation responses engage with this agenda.

These governance and sustainability themes are evoked in the next chapter that describes and discusses the key themes of adaptation politics.

## References

ADB (Asian Development Bank). (2014). *Technologies to support climate change adaptation in developing Asia*. Manilla: ADB.

Adger, W. N. (2003). Social capital, collective action, and adaptation to climate change. *Economic Geography, 79*(4), 387–400.

Adger, W. N., & Jordan, A. (Eds.). (2009). *Governance for sustainability*. Cambridge: Cambridge University Press.

Adger, W. N., Arnell, N. W., & Thompkins, E. L. (2005). Successful adaptation to climate change across scales. *Global Environmental Change, 15*, 77–86.

Allison, E. A. (2015). The spiritual significance of glaciers in an age of climate change. *WIREs Climate Change, 6*, 493–508.

Arrow, K., Bolin, B., Costanza, R., Dasgupta, P., Folke, C., Holling, C. S., et al. (1995). Economic growth, carrying capacity, and the environment. *Ecological Economics, 15*, 91–95.

Bacchi, C. L., & Goodwin, S. (2016). *Poststructural policy analysis: A guide to practice*. New York: Palgrave Macmillan.

Bailey, I., & Wilson, G. A. (2009). Theorising transitional pathways in response to climate change: Technocentrism, ecocentrism, and the carbon economy. *Environment and Planning a: Economy and Space, 41*, 2324–2341.

Barnett, J., & O'Neil, S. J. (2010). Maladaptation. *Global Environmental Change, 20*(2), 211–213.

Barnett, J., & O'Neill, S. J. (2013). Minimising the risk of maladaptation: A framework for analysis. In J. Palutikof, S. L. Boulter, A. J. Ash, M. Stafford Smith, M. L. Parry, M. Waschka, & D. Guitart (Eds.), *Climate adaptation futures*. Wiley Online Library. Hoboken: Wiley-Blackwell.

Barry, C., & Kirby, R. (2017). Scepticism about beneficiary pays: a critique. *Journal of Applied Philosophy, 34*(3), 285–300.

Barry, J. (1999). *Rethinking green politics: Nature, virtue and progress*. London: Sage.

Bauer, A., Feichtinger, J., & Steurer, R. (2012). The governance of climate change adaptation in 10 OECD countries: Challenges and approaches. *Journal of Environmental Planning Policy & Management, 14*(3), 297–304.

Blühdorn, I. (2019). Sustainability-post-sustainability-unsustainability. In T. Gabrielson, C. Hall, J. M. Meyer, & D. Schlosberg (Eds.), *The Oxford handbook of environmental political theory* (pp. 259–273). Oxford: Oxford University Press.

Boulding, K. E. (1978). *Ecodynamics: A new theory of societal evolution*. Beverly Hills: Sage.

Bulkeley, H., & Betsill, M. M. (2003). *Cities and climate change: Urban sustainability and global environmental governance*. London: Routledge.

Carter, N. (2009). *The politics of the environment: Ideas-activism-policy*. Cambridge: Cambridge University Press.

Daly, H. E. (1979). On thinking about future energy requirements. In C. T. Unseld, D. L. Sills, D. E. Morrison, & C. P. Wolf (Eds.), *Sociopolitical effects of energy use and policy* (pp. 232–240). Reports to the Sociopolitical Effects Resource Group, Risk and Impact Panel of the Committee on Nuclear and Alternative Energy Systems. Washington, DC: National Research Council.

Daly, H. E. (1990). Toward some operational principles of sustainable development. *Ecological Economics, 2*(1), 1–6.

Daly, H. E. (1996). *Beyond growth: The economics of sustainable development*. Boston: Beacon Press.

Daly, H. E. (2007). *Ecological economics and sustainable development, selected essays of Herman Daly*. Cheltenham: Edward Elgar.

Dewulf, A. (2013). Contrasting frames in policy debates on climate change adaptation. *WIREs Climate Change, 4*, 321–330.

Dobson, A. (1992). *Green political thought*. London: Unwin Hyman.

Dobson, A. (2016). *Environmental politics: A very short introduction*. Oxford: Oxford University Press.

Doherty, B. (2005). *Ideas and actions in the green movement*. London: Routledge.

Dryzek, J. S. (1987). *Rational ecology: Environment and political economy*. Oxford: Basil Blackwell.

Dryzek, J. S., Norgaard, R. B., & Sclosberg, D. (2013). Climate change and society: Approaches and responses. In J. S. Dryzek, R. B. Norgaard, & D. Schlosberg (Eds.), *The Oxford handbook of climate change and society* (Paperback ed., pp. 3–17). Oxford: Oxford University Press.

Duus-Otterström, G., & Jagers, S. (2011). Why (most) climate insurance schemes are a bad idea. *Environmental Politics, 20*(3), 322–339.

Eckersley, R. (1992). *Environmentalism and political theory: Toward an ecocentric approach*. Albany: SUNY Press.

Eckersley, R. (2004). *The green state: Rethinking democracy and sovereignty*. Cambridge: MIT Press.
Fankhauser, S., Smith, J. B., & Tol, R. S. J. (1999). Weathering climate change: Some simple rules to guide adaptation decisions. *Ecological Economics, 30*(1), 67–78.
Farber, D. A. (2013). Issues of scale in climate governance. In J. S. Dryzek, R. B. Norgaard, & D. Schlosberg (Eds.), *The Oxford handbook of climate change and society* (Paperback ed., pp. 479–503). Oxford: Oxford University Press.
Feenberg, A. (1999). *Questioning technology*. London: Routledge.
Fieldman, G. (2011). Neoliberalism, the production of vulnerability and the hobbled state: Systemic barriers to climate adaptation. *Climate and Development, 3*(2), 159–174.
Fischer, F. (1990). *Technocracy and the politics of expertise*. London: Sage.
Gabrielson, T., Hall, C., Meyer, J. M., & Schlosberg, D. (Eds.). (2019). *The Oxford handbook of environmental political theory*. Oxford: Oxford University Press.
Georgescu-Roegen, N. (1971). *The entropy law and the economic process*. Cambridge: Harvard University Press.
Gifford, R., Kormos, C., & McIntyre, A. (2011). Behavioral dimensions of climate change: Drivers, responses, barriers, and interventions. *WIREs Climate Change, 2,* 801–827. https://doi.org/10.1002/wcc.143.
Granberg, M. (Ed.). (2019). *Göteborgsregionen och klimatrisker*. Rapport 2019: 1. Karlstad: The Centre for Climate and Safety, Karlstad University.
Granberg, M., & Glover, L. (2014). Adaptation and maladaptation in Australian national climate change policy. *Journal of Environmental Policy & Planning, 16*(2), 147–159.
Granberg, M., & Nyberg, L. (2018). Climate change adaptation, city competitiveness and urban planning in the city of Karlstad, Sweden. In S. Moloney, H. Fünfgeld, & M. Granberg (Eds.), *Local action on climate change: Opportunities and constraints* (pp. 111–125). London and New York: Routledge.
Granberg, M., Bosomworth, K., Moloney, S., Kristianssen, A.-C., & Fünfgeld, H. (2019). Can regional-scale governance and planning support transformative adaptation? A study of two places. *Sustainability*, *11*(24), 6978. https://doi.org/10.3390/su11246978.
Grove, R. (1995). *Green imperialism*. Cambridge: Cambridge University Press.
Guha, R. (2000). *Environmentalism: A global history*. New York: Longman.
Hulme, M. (2009). *Why we disagree about climate change: Understanding controversy, inaction and opportunity*. Cambridge: Cambridge University Press.
Huseby, R. (2015). Should the beneficiaries pay? *Politics, Philosophy & Economics, 14*(2), 209–225.
IPCC. (2014). *Climate change 2014: Impacts, adaptation, and vulnerability*. Cambridge: Cambridge University Press.

Levin, K., Cashore, B., Bernstein, S., & Auld, G. (2012). Overcoming the tragedy of super wicked problems: Constraining our future selves to ameliorate global climate change. *Policy Sciences, 45*, 123–152.
Lovins, A. B. (1979). A target critics can't seem to get in their sights. In H. Nash (Ed.), *The energy controversy: Soft path questions and answers* (pp. 15–34). San Francisco: Friends of the Earth.
Lutz Warren, J. (2006). *Aldo Leopold's odyssey*. Washington: Island Press.
Magnan, A., Schipper, E. L. F., Burkett, M., Bharwani, S., Burton, I., Eriksen, S., et al. (2016). Addressing the risk of maladaptation to climate change. *Wiley Interdisciplinary Reviews: Climate Change, 7*(5), 646–665.
Marshall, G. R. (2015). Polycentricity, subsidiarity and adaptive efficiency. In *Ostrom workshop in political theory and policy analysis*. Bloomington: Indiana University.
Martínez-Alier, J., Pascual, U., Vivien, F.-D., & Zaccai, E. (2010). Sustainable de-growth: Mapping the context, criticisms and future prospects of an emergent paradigm. *Ecological Economics, 69*, 1741–1747.
McGormick, J. (1989). *Reclaiming paradise*. Bloomington: Indiana University Press.
Meadows, D. H., Randers, J., & Meadows, D. (2004). *The limits to growth: The 30-year update*. White River Junction: Chelsea Green Publishing.
Mettler, S., & SoRelle, M. (2014). Policy feedback theory. In P. A. Sabatier & C. M. Weible (Eds.), *Theories of the policy process* (3rd ed., pp. 151–182). Boulder: Westview Press.
Mills, E. (2005). Insurance in a climate of change. *Science, 309*(5737), 1040–1044.
Mol, A. P. J. (2016). The environmental nation state in decline. *Environmental Politics, 25*(1), 48–68.
Moloney, S., & Fünfgeld, H. (2015). Emergent processes of adaptive capacity building: Local government climate change alliances and networks in Melbourne. *Urban Climate, 14*(1), 30–40.
Moloney, S., Fünfgeld, H., & Granberg, M. (Eds.). (2018a). *Local action on climate change: Opportunities and constraints*. London and New York: Routledge.
Moloney, S., Fünfgeld, H., & Granberg, M. (2018b). Climate change responses the global to local scale: An overview. In S. Moloney, H. Fünfgeld, & M. Granberg (Eds.), *Local action on climate change: Opportunities and constraints* (pp. 1–16). London and New York: Routledge.
Nanda, V.P. (2006). The "good governance" concept revisited. *The ANNALS of the American Academy of Political and Social Science, 603*, 269–283.
O'Riordan, T. (1976). *Environmentalism*. London: Pion.
O'Riordan, T. (1999). Ecocentrism and technocentrism. In M. J. Smith (Ed.), *Thinking through the environment: A reader* (pp. 32–40). London: Routledge.

Olsson, D. (2018). *Conditions of 'sustainability': The case of climate change adaptation in Sweden.* Karlstad: Karlstad University Press.

Page, E. A. (2012). Give it up for climate change: A defence of the beneficiary pays principle. *International Theory, 4*(2), 300–330.

Parr, A. (2019). Capital, environmental degradation, and economic externalization. In T. Gabrielson, C. Hall, J. M. Meyer, & D. Schlosberg (Eds.), *The Oxford handbook of environmental political theory* (pp. 445–459). Oxford: Oxford University Press.

Patterson, J. J. (2018). *Adaptive cities? Institutional innovation under climate change: A global survey of 96 cities.* Institute for Environmental Studies (IVM), VU University Amsterdam, and Open University of The Netherlands.

Pelling, M. (2011). *Adaptation to climate change: From resilience to transformation.* London: Routledge.

Pepper, D. (1999). *Modern environmentalism: An introduction.* London: Routledge.

Piketty, T. (2014). *Capital in the twenty-first century.* Harvard: Harvard University Press.

Plumwood, V. (1993). *Feminism and the mastery of nature.* London: Routledge.

Ponting, C. (1992). *A green history of the world.* London: Penguin.

Renn, O. (2008). *Risk governance: Coping with uncertainty in a complex world.* London: Earthscan.

Richardson, D., & Rootes, C. (Eds.). (1995). *The green challenge: The development of green parties in Europe.* London: Routledge.

Rittel, H. W. J., & Webber, M. M. (1973). Dilemmas in a general theory of planning. *Policy Sciences, 4,* 155–169.

Sartori, G. (1987). *The theory of democracy revisited.* Chatham: Chatham House Publishers.

Schäfer, L., Kreft, S., & Warner, K. (2019) Exploring and managing adaptation frontiers with climate risk insurance. In R. Mechler, L. M. Bouwer, T. Schinko, S. Surminski, & J. Linnerooth-Bayer (Eds.), *Loss and damage from climate change: Concepts, methods and policy options* (pp 317–341). Cham: Springer Nature.

Scoones, I. (2016). The politics of sustainability and development. *Annual Review of Environment and Resources, 41*(1), 293–319.

Scoones, I., Leach, M., & Newell, P. (Eds.). (2015). *The politics of green transformations.* London: Routledge.

Sovacool, B. K. (2011). Hard and soft paths for climate change adaptation. *Climate Policy, 11*(4), 1177–1183.

Sovacool, B. K., & Brown, M. A. (2017). Scaling the policy response to climate change. *Policy and Society, 27,* 317–328.

Sovacool, B. K., & Linnér, B.-O. (2016). *The political economy of climate change adaptation.* Basingstoke: Palgrave Macmillan.

Surminski, S., Bouwer, L. M., & Linnerooth-Bayer, J. (2016). How insurance can support climate resilience. *Nature Climate Change, 6*, 333–334.

Surminski, S., & Eldridge, J. (2017). Flood insurance in England—An assessment of the current and newly proposed insurance scheme in the context of rising flood risk. *Journal of Flood Risk Management, 10,* 415–435.

Taylor, B. (2001). Earth and nature-based spirituality (Part I): From deep ecology to radical environmentalism. *Religion, 31,* 175–193.

Tompkins, E. L., Adger, W. N., Boyd, E., Nicholson-Cole, S. A., Weatherhead, E. K., & Arnell, N. W. (2010). Observed adaptation to climate change: UK evidence of transition to a well-adapting society. *Global Environmental Change, 20*(4), 627–635.

Tschakert, P., Ellis, N. R., Anderson, C., Kelly, A., & Obeng, J. (2019). One thousand ways to experience loss: A systematic analysis of climate-related intangible harm from around the world. *Global Environmental Change, 55*, 58–72.

Turnbull, N., & Hoppe, R. (2019). Problematizing 'wickedness': A critique of the wicked problems concept, from philosophy to practice. *Policy and Society, 38*(2), 315–337. https://doi.org/10.1080/14494035.2018.1488796.

UNFCCC. (2006). *Technologies for adaptation to climate change.* Bonn: UNFCCC Secretariat.

Unwin, K., & Jordan, A. (2008). Does public policy support or undermine climate change adaptation? Exploring policy interplay across different scales of governance. *Global Environmental Change, 18*(1), 180–191.

Wackernagel, M., Schulz, N. B., Deumling, D., Callejas Linares, A., Jenkins, M., Kapos, V., et al. (2002). Tracking the ecological overshoot of the human economy. *Proceedings of the National Academy of Science, 99*(14), 9266–9271.

WCED. (1987). *Our common future.* New York: The United Nations.

CHAPTER 5

# Political Themes in Adaptation

**Abstract** This chapter describes and discusses the key themes in adaptation politics: Equity, decision-making, power, Nature, gender and commons management. A further section covers what are described as 'implicit' (or embodied) politics in two forms of adaptation, resilience (and its association with neoliberalism) and the politics in socio-ecological transformations. These themes constitute the elements that are commonly debated within the adaptation discourse over specific issues, places and processes. Opening the chapter is an identification of two cross-cutting themes: 1) Adaptation as an agent of power that distributes the costs and benefits of climate change impacts across the social and natural realms and 2) adaptation as a political process that both shapes, and is shaped by, the interests and representations of stakeholders (including the interests of the natural world).

**Keywords** Decision making · Participation · Ecology · Equity · Fairness · Gender · Power · Nature

A core set of political themes feature in adaptation politics studies; identifying these provides a guide to the expressions of their key political ideas. Six prominent themes in the relevant scholarship are overviewed in this



L. Glover and M. Granberg,
*The Politics of Adapting to Climate Change*,
https://doi.org/10.1007/978-3-030-46205-5_5

chapter: Equity and fairness; decision making, participation and representation; power and influence; Nature and ecology; gender; and governance: lessons from commons management. Additional to these explicit themes are the 'implicit' politics that are examined at the chapter's end.

Oftentimes in adaptation scholarship, two or more of these themes are considered in conjunction, overlap or intersect in some way. Two such political cross-cutting justice topics emerge strongly from these themes:

- Distributional (or allocative) issues arising from adaptation responses' influence on socio-economic factors, and
- Participatory issues involving the processes, institutions and implications of citizen and stakeholder engagement.

Paavola and Adger (2006) differentiated between distributive and participatory justice in considering fairness in adaptation. Jagers and Duus-Otterström (2008) argued that issues of distributive justice arise not only in relation to those who take adaptive action but also in relation to recipients of benefits. Thomas and Twyman (2005) considered that although fairness in climate change relates mostly to impacts, vulnerability and adaptation measures, there are also distributive and procedural aspects to climate change justice. As discussed below, most of the work on these themes have had a strongly progressive outlook (e.g., favouring greater fairness, promoting democratic goals and seeking nature conservation), although this is not necessarily acknowledged.

A third notion of justice evident in adaptation studies relating to justice and capabilities, as widely promoted by Amartya Sen (1999), wherein justice involves meeting the (basic) needs of individuals (and societies) and capabilities that enable social 'flourishing' (Schlosberg 2011). Each of these six explicit political themes is reviewed, in turn, below.

## Equity and Fairness

Equity issues cover the pre-existing socio-economic conditions (and how these influence risk exposure, especially vulnerability), together with the known or forecast socio-economic effects of adaptation measures, to which are added the effect of these factors on the capacity/effectiveness of adaptation measures (Jagers et al. 2009; Sowers et al. 2011; Baja and Granberg 2018). Equality and equity are largely considered as outcomes or desiderata, although conceptually the splitting of means and ends creates some logical and philosophical problems.

Equity, as one of classical philosophical tropes, is understood in many different ways. One interpretation is based in ideas about justice, in that equal cases receive equal treatment, but it needs to be recognised that equity and equality are not identical. According to Baja and Granberg (2018, p. 126):

> Equality is treating everyone the same. It makes the assumption that if all people are given the same resources and options, that they are being treated fairly. However, that only works if everyone has the same set of circumstances. Equity involves understanding the history and context and ensuring people receive the resources they need in proportion to what they have been denied historically.

This clearly highlights the need for a longer historical perspective and Mearns and Norton (2010) state that the causes and consequences of climate change are intertwined with historical patterns of inequality. In the 'New York City Panel on Climate Change 2019 Report' a full chapter was dedicated to equity issues related to climate change adaptation (Foster et al. 2019). They highlighted the need to include a full analysis of pre-existing social and economic challenges in specific communities in assessing vulnerabilities and in designing and implementing adaptation measures.

Justifications for taking equity into account usually take either ethical (i.e., pursuit of justice) or normative forms (i.e., pursuit of policy efficacy). Some of the debate over equity and adaptation takes into consideration pre-existing social inequities; these are more than socio-economic and can include inequities over economic opportunities, access to political power and political participation (Reeves 2005; Wilkinson 2012; Baja and Granberg 2018). Historians and political philosophers have repeatedly observed that beyond certain extents of inequality and unfairness in resource distributions, opportunities and access to political power, there is social breakdown wherein governments and ruling classes loose legitimacy, authority and positions of power. Equity is often listed in adaptation policy documents as a goal but in the absence of specific instruments to fulfil such intentions, which is frequently the case, equity is merely an empty and misleading trope in these circumstances.

There are also other interpretations of equity and fairness. Adger et al. (2017), identify three moral positions dealing directly with fairness: those based on ability/need/entitlement, those based on fairness in burdens and those based on protection from harm. Hence, there are many approaches to egalitarian goals, including one that appears frequently in adaptation studies, namely giving highest priority to those in greatest need (i.e., the most vulnerable) (Paavola and Adger 2006). Another interpretation often employed in environmental studies is intergenerational equity, which deals with the obligations of living generations towards future generations.

A confounding philosophical issue in dealing with equity is the difference between justice as a uniform or universal state ('cosmopolitanism') and a conditional one (sometimes termed 'communitarianism') (see, Waldron 1992), i.e., dependent on local circumstances and needs (aka, universalism versus particularism). Although the term 'equity' is often used in adaptation studies considering social factors, accompanying articulations of its meaning are far less common. It appears that most adaptation studies dealing with equity implicitly adopt a conditional approach, given that attention is paid to specific aspects of (social) vulnerability, including sensitivity to risk, risk exposure and capacity to cope and respond.

Given the moral claims of equity and fairness, any other rationalisations for pursuing equity in adaptation seems unnecessary. However, such claims in isolation are not always sufficiently persuasive to motivate appropriate public policy in any realm, making other supplementary rationalisations needed for application to adaptation. Disregarding any reasonable notion of equity and fairness can lead to highly iniquitous and unjust outcomes. Following a purely market-driven approach to adaptation simply results in leaving every individual, household and business to their own devices in efforts to avoid losses from climate change (Granberg and Glover 2014). As a result, the level of adaptation carried out is determined by those with the resources available for investing in adaptation. Some believe this this is close to what is largely occurring, such as Grove (2016, p. 173):

> There are winners and losers in any adaptive action, but the game is usually rigged from the start; the people who adapt are usually those with the social, political, and financial resources to do so.

Those left behind in such preparations are those without resources and they remain vulnerable to climate change impacts (Baja and Granberg 2018). Not only will the costs of these impacts worsen the circumstances of those least well-off, but his inequity will be become more marked as the welfare of those able to undertake adaptation will be relatively increased. Further, in capitalist societies, those with more resources in post-disaster circumstances are in a position to take advantage of those made worse-off. Reconstruction of New Orleans after the 2005 Hurricane Katrina is something of a poster child for post-disaster opportunism, putting aside the obvious expressions of US federal government crony capitalism for favoured corporations (e.g., Dreier 2006), speculators rapidly bought housing properties in the aftermath to profit from the housing shortage and federal funding and authorities took advantage of the crisis to promote a raft of neoliberal policies to promote private interests, (i.e., the 'disaster capitalism' mentioned in Chapter 3), (also see, Quigley 2008; Adams 2012). In effect, disasters constitute the original economic 'disruptors', creating new opportunities for profiteering and exploitation of the disadvantaged. Therefore, the case for purposefully pursuing equity is to prevent the inherent historical biases in socio-economic conditions from increasing these disparities when responding to climate change impacts (Reeves 2005).

Adaptation has been implicitly tied to sustainable development discourse from the outset, a discourse carrying an implicit and normative position on reducing socio-economic inequality (WCED 1987; Olsson 2018). However, the adaptation discourse has not expressed a wider interest in equity issues or become strongly linked to development policy and research until comparatively recently (Abeygunawardena et al. 2009; Adger et al. 2006; Huq et al. 2006; McGray et al. 2007). While most attention is directed towards adverse effects on equity, climate change impacts also have the potential to produce beneficial effects that also need to be considered in equity aspects of adaptation (Ruhl 2012). It also worth remembering that climate change will cause new inequities.

In line with the reasoning above, there has been a realisation in the climate change debate that poor people in poor nations will bear the brunt of negative climate change impacts (Ribot 1995; IPCC 2001). Women will be disproportionately harmfully affected (Denton 2002; Terry 2009) and indigenous peoples are particularly vulnerable to climate impacts. Other highly vulnerable social groups include young children and the elderly. Communities dependent on natural resources in developing countries are recognised as being of highest concern (Tol et al. 2004).

Heedfulness over equity in adaptation, however, should not be restricted to developing nations; vulnerabilities to climate change hazards cross the global North-South divide. Assumptions that developed nations with their low vulnerabilities are without any equity concerns over adaptation, are misleading (e.g., consider Hurricane Katrina in the US, Australia's Millennium drought and extreme temperatures in Southern Europe in the early 2000s). Within wealthy nations there are regions, communities and households that are highly vulnerable, partly as a consequence of historical economic, political and other disadvantages. As an example, in New York City this has been addressed through community-based analysis of adaptation approaches (Foster et al. 2019). This included investigating spatial patterns of social vulnerability to climate change, case studies in socially and economically disadvantaged communities, examining community-based adaptation planning efforts and analysing current practices for opportunities to incorporate equity measures.

In the international context, the pressing equity issue regarding adaptation is the transfer of resources and other support from developed nations to lesser developed nations (Sovacool and Linnér 2016). Many basic equity questions can be posed. What level of support should be provided? What levels of burden should the developed nations carry individually? How should this support be allocated to individual lesser developed nations and what allocations would be equitable? Few would dispute that the developed nations that have reaped the benefits of developed that resulted in GHG emissions, so that they would carry the moral responsibility for assisting developing nations to deal with the costs of climate change impacts. This principle is enshrined in the UNFCCC, covered by several paragraphs in Article 4 of the Framework Convention (UNFCCC 1992).

It is clear that climate change will have a range of social effects that will either add or subtract from pre-existing social inequalities. Responding to this public policy challenge by states will therefore influence the circumstances of those affected, as will decisions not to respond. There is, however, no blueprint for achieving fair and equitable outcomes from undertaking adaptation activities, but there are a number of measures through which the outcomes of public policy on equity can be assessed.

## Decision Making, Participation and Representation

Decision making processes are the 'engine room' of adaptation governance and thus are the locus of the procedural aspects of equity and fairness; public-sector adaptation decision making in particular has attracted political scholarship as it entails government institutions and associated political processes (Dessai et al. 2009; Moloney et al. 2018a). As is known, the bulk of adaptation decisions has, and will, comprise those taken by individuals, households and businesses. Although individual decision making appears outside of what might be conventionally viewed as 'politics', this does not mean that these have been taken without outside influences. Adaptation responses to complex, widespread or multi-faceted climate hazards are rarely confined to the domain of a single scale or actor; in practice, there will be mixture of individual and collective responses across several tiers of government.

As with the case for greater distributional equity, the reasoning for procedural justice tends to be based either on efficacy or moral claims. Producing decisions acceptable to actors that leave the authority of decision makers undiminished is a pillar of the efficacy arguments (Benhabib 1994). Paavola and Adger (2006, p. 601) argued:

> Distributive justice is unlikely to be sufficient for climate justice—procedural justice is needed to underpin the legitimacy of climate change regime.

A number of studies support wider stakeholder participation on the grounds that it produces more credible, salient and effective policies with greater prospects for success by overcoming some of recognised barriers (e.g., Smith et al. 2009; Hegger and Dieperink 2016).

Procedural equity issues involve moral choices and evoke a range of opinions evoked by differences in political values, as with the issues of distributional equity and fairness (Paavola 2005). Collective decisions require that such differences are resolved, compromises reached and accommodations made between actors with different values for governance to retain its legitimacy. Procedural justice features decision making processes that are deemed to be fair (by the participants), accessible (meaning inclusive of all those involved) and transparent (typically meaning that decision makers are accountable to participants). Recognition has become a major feature of the calls for greater procedural justice following the wider social movement (Fraser 2001). A number of writers considering procedural justice identify recognition, participation and the distribution of power as key determinants in decision making activities (Fraser 2001; Paavola 2005).

Adaptive capacity of individuals and communities can be influenced by governance factors, such as resource access, political relationships and potential for participation, available infrastructure and general institutional responsiveness (Baja and Granberg 2018; Ensor et al. 2018). Greater equity in the distribution of power and resource access improves adaptive capacity across levels (local, regional, national and international), an outcome requiring suitable institutions and arrangements. As Paavola and Adger (2006, p. 601) state:

> Procedural justice fosters legitimacy because it assures those whose interests are not endorsed by a particular decision that their interests have been considered and that they have a chance to count in other decisions.

One manifestation of wider engagement is the UNDP community-based adaptation programme (UNDP 2010) that uses community participation in planning, implementing and monitoring activities, as well as other initiatives (Ebi and Semenza 2008; Reid et al. 2009; Dumaru 2010). Beginning as largely an NGO initiative, community-based adaptation is being taken up by bilateral and multinational agencies in efforts to 'mainstream' the practice (Reid and Huq 2014; Reid 2016).

Going beyond the goals for greater efficiency and effectiveness in the formulation and implementation processes are the cases for greater inclusion based on explicit democratic or other philosophical or ethical principles. UNESCO (2017) released the Declaration of Ethical Principles in Relation to Climate Change that includes Art. 4 on equity and justice which states that 'Justice in relation to climate change requires fair treatment and meaningful involvement of all people'. It explicitly calls for inclusion with: '… the poorest and most vulnerable people …' and that '… measures should take into account the contribution of women in decision-making.'

Mikulewicz (2018, p. 21) makes the point that these kinds of participatory efforts should not be equated with democratisation efforts, as:

> Interventions based on local institutions do not generally engage with issues of power and democracy, despite being founded on the principle of involving local people in the design, implementation and, less frequently, monitoring of adaptation or development interventions.

Mikulewicz (2018) argues that top-down and bottom-up approaches have usually overlooked the historical 'political roots' of vulnerability when considering natural hazards and socio-economic causes of vulnerability.

Adger et al. (2011) and Hale et al. (2013) identified issues of under-representation of vulnerable and often disenfranchised social groups and the interests of ecological protection in climate change decision making as ethical concerns. Tanner and Allouche (2011) characterise the dominant analytical approach to decision making as featuring collective action, rational choice and rent seeking and offer 'new political economy' in its place, featuring ideological drivers, incentives and power relations.

## Power and Influence

Political power is of interest to those seeking to examine, measure, analyse and interpret the governance of adaptation, especially through the functioning of institutions and is, in all likelihood, the most commonly used explanatory factor in descriptions and analyses of the politics of adaptation (Moloney et al. 2018b). In an extensive literature of adaptation governance covering knowledge and power comprising 1132 articles, Vink et al. (2013) indicated something of the extent of interest in these themes. Power can have a multitude of interpretations and expressions (e.g., Lukes 2005). Fazey et al. (2016) recognise power in the adaptation discourse in two broad forms:

1. 'Power over others' (comprising formal, persuasive and unconscious forms), and
2. 'Normative pathways' (i.e., the effects of discourses and practices).

Most work dealing with power in adaptation adopts a formal and restricted understanding of the concept (i.e., based in governments) and even studies covering so-called 'unorganised' power usually consider this through the lens of access and/or influence to formal power (Vink et al. 2013).

Political power is a factor in understanding the predetermining effects of the mediation of biophysical change by political and social factors (e.g., social vulnerability is a social and ecological construct). For the most part, the adaptation discourse focuses on representation with views that this necessitates exchanges between public, community and citizenry interests and the policy process, featuring mutual responsiveness and accountability. Policy framework processes have traditionally had no place for political considerations and have been poorly equipped to deal with political circumstances (Rickards 2010; Pelling 2011).

Inequalities in power and influence can be lessened by power-sharing in decision making (Beetham 2013, 2018). As described above, the numerous advocates of greater inclusiveness in decision making (especially in the form of community-based adaptation) promote this both for its normative and other positive benefits (Reid et al. 2009; Reid 2016; Kirkby et al. 2017). A far smaller number of contributors recognise the limitations imposed by the absence or lack of agency for some stakeholders, whereby political representation can fail to overcome pre-existing differences in political (and economic) power (Mansuri and Rao 2012; Ensor et al. 2018). Kirkby et al. (2017, p. 582) state that:

> In practice, participatory processes have often been reduced to a means for powerful institutional actors to legitimate and build public acceptance for pre-determined agendas, policies and interventions.

Accordingly, stakeholder engagement is now recognised as an essential aspect of adaptation formulation and implementation but this institutional realm entails an array of political interests, conflicts and uncertain outcomes. Nightingale (2017, p. 12) offers:

> If power and politics reshape the purpose of adaptation efforts, then adaptation becomes about adjusting to entangled socio-political contestations, biophysical change, livelihood desires, struggles for authority to govern change, and desires for social and political recognition by those promoting programs and recipients of them.

Hence, despite widespread promotion of increased stakeholder representation in adaptation institutions, current practice has consistently failed to meet such expectations. Agrawal and Perrin's (2008) review of National Adaptation Programmes of Action projects, for instance, found only 20 of 173 projects had local-level institutions as partners or agents. While wider stakeholder participation may satisfy formal democratic ideals for institutional practices, critical concerns arise from the risks of 'tyranny' with community-based adaptation (Dodman and Mitlin 2013). In other words, asymmetric power relations entail that those with pre-existing advantages can readily exploit others with less resources in decision making regardless of the scale or jurisdictional realm.

Some scholars have considered the barriers to wider participation in adaptation research, planning and practices and Kirkby et al. (2017) identify several problems with achieving meaningful participation: most adaptation relies on (technical and) scientific knowledge and outside experts that are 'privileged' over local, indigenous knowledge; participatory processes tend to reflect existing power relationships and social hierarchies; and that wider participation is not always warranted and can yield divisive and divergent views. As many studies have shown, political power—or more importantly, asymmetries in political power, are historically interwoven in the socio-economic fabric of social life (Baja and Granberg 2018). Public policy interventions may be successful in neutralising or counteracting such inequalities within the realm of specific initiatives, but rarely produce widespread social changes. Furthermore, such outcomes are necessitated by decision making processes that can counteract the power asymmetries within decision making that typically requires those with greater power to act against their own interests.

## Nature and Ecology

Fundamentally, adaptation is a social response to a problem of an anthropogenically-disrupted ecological system. Accordingly, ecology occupies a central and complicated place in adaptation policy and research. As the IPCC (2014, p. 840) states:

> Climate change is altering ecological systems, biodiversity, genetic resources, and the benefits derived with ecosystem services … Climate change is inducing shifts in habitats that often cannot be followed by species … leading to changed ecosystems, to local and global extinctions, and to the permanent loss of unique combinations of genes.

At least three major approaches to Nature and ecology in adaptation scholarship and practice can be identified. *Firstly*, there is ecology as a component in the socio-ecological construct. Much of this work relates to the fields of risk and hazard management, notably the effect of climate change on natural hazards and climate-related extremes. Natural processes and ecological phenomena (see, e.g., IPCC 2014):

- Are the major source of detrimental climate change impacts and risks, thereby creating the need for social adaptation (i.e., an influence on the determinants of biophysical vulnerability), and
- Influence pre-existing or current adaptability/adaptive capacities (i.e., an influence on the determinants of adaptation).

Nature, in this setting, is exogenous to the social realm and is a carrier of actual or potential costs and benefits that are mediated by policy and other social responses.

*Secondly*, there is the wide field of adaptation research and policies for natural resource management for socio-economic goals, applied at the full range of possible scales. Of particular concern are the provisioning, regulating and servicing 'ecosystem services', that include food and fibre production, potable water, pollination, disease control and nutrient cycling. According to the IPCC (2014, p. 841):

> Natural systems underpin human livelihoods, health, welfare, food security, and prosperity. Vital ecosystem services that need to be maintained include provisioning services such as food, fiber, and potable water supply; regulating services such as climate regulation, pollination, disease control, and flood control; and supporting services such as primary production and nutrient cycling … or instance, healthy coastal wetlands and coral reefs can help to protect against storm surges and rising sea levels … while the maintenance of wetlands and green spaces can control runoff and flooding associated with increases in precipitation.

Thereby making protecting these natural systems and resources is an imperative in adaptation responses for human flourishing (at the least)—if not for outright survival, if climate change takes a catastrophic turn (IPCC 2014).

Although most adaptive responses have involved technological and engineering measures, use of ecosystem-based measures is increasing, such as ecological restoration, afforestation, ecological corridors and community-based natural resource management (IPCC 2014). Colls et al.

(2009, p. 1) have highlighted advocacy of 'ecosystem-based adaptation' as an important development in the field, that:

> … integrates the use of biodiversity and ecosystem services into an overall strategy to help people adapt to the adverse impacts of climate change. It includes the sustainable management, conservation and restoration of ecosystems to provide services that help people adapt to both current climate variability, and climate change. Ecosystem-based adaptation contributes to reducing vulnerability and increasing resilience to both climate and non-climate risks and provides multiple benefits to society and the environment.

This stands in contrast to technology-based strategies and offers a diverse, lower-cost and effective range of options amenable for application by poorer communities dependent of natural ecosystem services and products (Vignola et al. 2009; Andrade Pérez et al. 2010; Munang et al. 2013). Essentially, natural elements here have a utilitarian cast and adaptation is legitimated using anthropogenic goals, i.e., their value to society. Nature and ecosystems are accorded an extrinsic value.

*Thirdly*, the field of investigations and public policy concerned with protecting natural values from climate change impacts; i.e., conservation in broad terms (IPCC 2014). Much of this research work is descriptive and analytical, having the goal of understanding climate change impacts on species, species assemblages, ecosystems and ecosystem processes. An increasing portion is applied research with the goal of assisting policy formulation for protecting existing values or restoring degraded values through adaptation measures (Heller and Zavaleta 2009; Mawdsley et al. 2009; Vignola et al. 2009). Peterson (2009) points to the ecological limits to adaptation in the face of:

- Extensive and on-going human modification of natural ecosystems undermining the scale of reliability of ecosystem services, and
- Declining ecosystem capacities for self-regulation.

Adaptation, therefore, should aim to reduce these limits in order to build ecological resilience, Peterson (2009) argues.

There are important strategic differences and conflicting priorities in policy directions for ecological protection; implementation barriers and opportunities give rise to the usual array of institutional and policy factors

(West et al. 2009). A distinguishing feature here is the recognition of the values of (non-human) entities and towards which there are moral obligations, i.e., Nature is valued intrinsically. Most of this activity occurs as Nature conservation, although there is an increasing body of work that combines Nature conservation and development goals that can be classified as socio-ecological in character.

## Gender

Development theory and research has highlighted gender differences—specifically the disadvantages suffered by women—since the 1970s, including making gender a component of sustainable development (see, e.g., Harcourt 1994). Disaster and natural hazard scholarship have been slower in addressing gender, only gaining momentum in the 1990s (see, Enarson et al. 2006). Demetriades and Esplen (2008) pointed out the research on gender and agriculture, forestry, livelihoods, migration and water management that could inform adaptation and gender studies. Gender, however, was largely neglected climate change adaptation discourse and there were many calls for its inclusion (see, e.g., Denton 2002; Lambrou and Piana 2005; Nelson et al. 2002). Terry (2009) is of the view that this was the case until 2002. IPCC's review of available knowledge in 2001 offered nothing specific on the issue; it appears in its fourth assessment in 2007 (IPCC 2007) and is covered in some detail in the 2014 assessment (IPCC 2014) and there has been considerable research into gender issues over the last decade or so. Unevenness in thematic coverage is to be expected but some critical gender issues have been neglected, such as health impacts (see Preet et al. 2010). This historical sequence is informative as sustainable development and disaster studies, and critiques of them by feminist scholars and others, have informed much of the subsequent adaptation discourse—particularly the 'gendering development' and 'mainstreaming gender' development debates.

Gender differences in climate change adaptation are now explored by research and scholarship, albeit highly varied in extent and range of themes (Arora-Jonsson, 2011; Kinnvall and Rydström 2019; Wester and Doma Lama 2019). Most effort has been directed at describing the vulnerabilities of women from the effects of climate and climate-related impacts, including their relationship with, and dependency on, the natural environment. Other themes covered are the effects of adaptation measures taken on women (especially disadvantages), the particular perceptions of women towards impact risks, the under-representation of women and women's interests in policy and planning fora (especially at the international level). Two rationales for greater recognition, inclusiveness and identifying benefits to women in this discourse appear frequently; ethical cases based around gender equity and social justice goals (and eliminating disadvantage) and those promoting efficacy gains, pointing to improved preparation, better planning outcomes, lower social and economic losses, more effective disaster recovery and the like.

Those societies, locations and nations most vulnerable to climate disasters and severe weather events will be most vulnerable to climate change; they will also likely to be less prepared, have less capacity to respond, will suffer the greatest socio-economic losses and disruptions relative to their socio-economic status and will face the greatest challenges in recovery and adapting to changed conditions (Lopez-Carresi et al. 2013). Women will be most affected by climate change impacts and the disadvantages suffered by women can be worsened by the effects of climate change (such as the extent of inequality) (Gaillard et al. 2017). These vulnerabilities are a consequence of pre-existing vulnerabilities to climate-related disasters and to climatic variability; i.e., the influence of social (in the broadest sense) and environmental context. Analysis of these vulnerabilities has explored the multiple dimensions and expressions of poverty beyond the absence of material wealth, including ill-health, lack of time, insecurity, uncertain identity and legal standing, weak or absent political representation and engagement in decision making. Many rural women face the multiple burdens of caring for households, children and the elderly, as well as generating income and foodstuffs. As Demetriades and Esplen describe (2008, p. 25):

> Each dimension intersects with gender and other forms of inequality (class, age, 'race'/ethnicity and sexuality), with implications for women's and men's vulnerability to – and capacity to adapt to – climate change.

Women's vulnerabilities are revealed at the individual and household level, but there are overarching cultural and historical factors giving rise to this condition, such as social class, marital status, socio-economic status, education levels and household responsibilities. Denton explains (2002, pp. 16–17; see also, Kelman 2020):

> Ineffective, gender-blind policies, and entrenched patriarchal traditions, are to a large extent responsible for the numerous constraints women face in transport, access to land, income, and other resources, agricultural practices, education, health services, credit facilities, and a litany of structural, technological, and cultural barriers.

Research into women and adaptation has identified their role in disaster recovery and restoration and, to a lesser extent, in adaptive practices. Women's knowledge, their roles in households, communities and in local economies are critical in adaptive efforts.

Women's under-representation in decision making in policy and planning expresses another form of disadvantage. Skutsch (2002) observes that neither the UNFCCC and the Kyoto Protocol use the words 'gender' or 'women', nor do the words 'poverty' or 'deprivation' occur either. A number of studies describe the exclusion of women from adaptation policy and planning in local, national and international institutions (e.g., Skutsch 2002; Villagrasa 2002).

UNEP recommends a number of actions to address the adaptation needs of women, including targeted programmes in agriculture, food security, natural resource management and rangelands, addressing women's resources access, ownership and control, investing in 'green' technologies, disseminating climate change data and information, increased participation and meeting education/training needs (Nellemann et al. 2011). While the UNEP has a more governmental approach, the UNDP promotes a community-based guidebook for action (UNDP 2010). NGOs have also been engaged in the issue: Oxfam published a training manual on gender and disaster risk reduction (Ciampi et al. 2011), as did the IUCN and UNDP (Aguilar 2009). These initiatives represent the concept of gender mainstreaming in action. 'Gender mainstreaming' has been evoked as a policy tool, namely assessing the gender implications of policies, programmes and actions with the goal that subsequent benefits are equally distributed between men and women (UNDP 2010).

In many respects, the treatment of gender has evolved since the early 2000s; broad-brushed associations between women and vulnerability have been superseded by studies dealing with more politically explicit concerns, particularly with structural issues, notably with political power and the causes of gender inequality. Jerneck (2018) employs the concepts of 'gender imperative' for the place of gender in social relationships, 'gender order' for describing how power relationships and institutions shape notions of masculinity and femininity and 'gender regime' for how these institutions function in specific circumstances. Explicitly, some question what they interpret as the earlier depiction of women as 'vulnerable victims' of future impacts (and previous disasters) and as 'virtuous enablers' of adaptive responses and restorers following disasters (Arora-Jonsson 2011; Demetriades and Esplen 2008; Tschakert and Machado 2012). These concerns have several dimensions. One is the misleading representation of diverse cultures and social settings in homogenous terms that decontextualizes women's identities and local circumstances. Another is that the narratives of 'helpless' and 'voiceless' women are disempowering and downplays the critical place of poverty in understanding women's vulnerability (Gaillard et al. 2017). Resurrección also promotes an appreciation of gender in context (2013, p. 41):

> A climate change policy regime will therefore benefit less from political imaginaries of women and environment ties, but from flexible readings of life on the ground, or in short, a stronger and more complex social analysis of climate, environment, power and people that informs response and action.

## Governance: Lessons from Commons Management

Commons management bridges the topics of governance and sustainability because of the political interest in its institutions and the environmental management interest in its successes in achieving environmentally sustainable resource management of renewable resources (Ostrom 1990; Dietz et al. 2003; Araral 2014; Forsyth and Johnson 2014; Wall 2017). Here, we concentrate on the governance and institutional aspects of commons, as these have attracted the greater scholarly attention related to adaptation politics. At its core, managing commons through collective efforts has a behavioural element (based on trust and reciprocity), namely that coordinated efforts between individuals for common goals can achieve outcomes that individual (or corporate) self-interest or state powers cannot. Effective commons management of natural resource systems has come to be associated with communities of higher resilience to climate change impacts, thereby constituting an effective adaptive strategy and attracting advocacy from scholars and official bodies. Commons management is a form of community-based adaptation and in the cases of indigenous peoples, typically employs traditional ecological knowledge (TEK) (see, e.g., Berkes 2012).

Although the 'climate-resilient pathways' model conceptually nests the practical (behaviour and technical responses) within the political (systems and structures) within the personal (beliefs, values and paradigms) (see, IPCC 2014, Ch. 20), for commons management these elements are mingled and interwoven with culture. However, there have been many efforts to analyse commons management empirically to understand the elements of its successful environmental management, of which Elinor Ostrom's (1990) study is the most highly regarded (although the IPCC has yet to give proper account to commons concepts). Ostrom's (1990) work on commons management institutions has attracted wide scholarly attention and is foundational in this field of inquiry (and especially well known for its refutation of the discredited 'tragedy of the commons' thesis).

Ostrom's (1990) principles for successful commons management have been much reviewed and applied in the adaptation scholarship; her study of local natural resource management schemes matches that at which many consider that adaptation efforts should be directed. In a sense, the foundational principles are that commons are defined, potential users can be excluded, that the resources are worth the effort of managing collectively, that rule makers have autonomy and that negotiations are possible over the rules put in place. Other principles focus on pragmatic institutional issues, namely that commons are monitored (including their usage and condition), sanctions for abusers must be graduated, conflict resolution is simple and straightforward, commons institutions are recognised as legitimate by higher authorities and commons management works best when nested in larger networks.

There have also been critical efforts to modify, enhance and extend Ostrom's ideas (see, e.g., Stern 2011; Araral 2014). One criticism is that Ostrom was overly focused on bottom-up measures and disapproving on higher-scale activities, whereas these can clearly be effective in environmental management. Although Wall (2017) considers Ostrom to be a prominent political ecologist, especially for her contributions of recognising the importance of democratic decision making in commons management, another critique is that Ostrom was largely disinterested in issues of power. Ostrom's theories are based on rational choice theory and some find this problematic, in part because this made any subsequent analysis reductionist. Critics argued that this economic abstraction produces outcomes divorced from local context and cultural elements

(Forsyth and Johnson 2014). A specific criticism is that the management systems described by Ostrom and her followers were developed to fulfil cultural goals, rather than resource management goals, per se (Forsyth and Johnson 2014).

Scholars working on the problems of institutions for adaptation have drawn on commons management concepts and practices. Dietz et al. (2003), for example, identified the governance requirements for complex systems. These were: (1) Providing information about the resource stocks and flows and human interactions with these systems, (2) dealing with conflicts between actors, (3) using formal and informal rules to ensure compliance, (4) providing infrastructure and (5) creating institutions to enable adaptation. Thompkins and Adger (2004) examined natural resource management for conditions that improved resilience and found that collective action appears to work best at the small-to medium scales, that successful care of the commons tends to be associated with more equitable distribution of the collective benefits amongst the group and alternatives are found when existing institutions are found to be failing (i.e., learning).

In effect, the adaptation discourse has in some respects turned back to older, and in some cases, ancient institutions for environmental management for lessons for contemporary managers and policymakers. Studies of commons management have also shown commons disruption by globalisation, conventional economic development, exposure to market capitalism and other forces (Berkes 2008). More typically, interest in commons management has focused on explaining differences in vulnerability/preparedness and in building resilience; commons management has also been evoked as having elements that can manage natural resource systems sustainably. Further, it has also served to instruct on why the privatisation of commons leads to over-exploitation of ecosystem goods and services. Commons management, however, struggles to provide a coherent theory for social change, i.e., transformative adaptation.

## 'Implicit Politics' in Resilience and Transformation

Social science has long drawn on studies of the natural sciences for building social theories and concepts, framing, models and paradigms, including those for adaptation, disaster response, development and environmental sustainability. Anthropomorphizing from ecology to socio-ecological systems provoked a number of criticisms of a political character (see, Chapter 2); here two key political themes in resilience and transformation, as drawn from socio-ecological foundations, are examined.

## Resilience as Implicit Neoliberalism

Resilience in social systems to climate change, in the orthodox perspective, is the ability to maintain existing social structures and functions in the face of disturbance through changes to relevant organisations and institutions (cf., Holling 1973; UNISDR 2009; Pelling 2011). Progressive change is possible through resilience measures within specific organisations or institutions, but this does not translate to more widespread reform; as noted above, these adaptations can enable environmentally unsustainable or socially unjust practices to continue. Features of this resilience model mark it as politically conservative or favouring the continuation of extant socio-economic (and socio-ecological) conditions. Generally, there is no place for treating values, with the associated shortcoming regarding the role and influence of political power in decision making; partly, this is explained through the model's narrowly formalist focus on decision making outcomes (rather than processes) (cf., Vale 2014). Consistent with the managerialism and technological optimism of resilience activities, political phenomena (including its cultural contexts) do not fit readily into these explanations. From the perspective of political analysis, a problem with the proponents of resilience models has been a lack of engagement with social theory.

Resilience has become regarded by some critics as a neoliberal phenomenon across many of the fields where it is used, and of relevance to our inquiry, including that of development (and sustainable development) and crisis responses (e.g., Reid 2013). Ideologically and functionally, the use and applications of resilience in adaptation have also been described as neoliberal. As such, these are stronger claims than those of Pelling (2011), along with others, who argue that resilience is conservative and a largely inadequate adaptation response, primarily because it addresses the symptoms of vulnerability rather than causes.

Examining the application and usage of resilience, Joseph (2013) posits that it is largely a neoliberal model of governance. Joseph finds that neoliberal governance occurs where that although the state withdraws form the social sphere and encourages individuals' activities, it also intervenes in civil society and making new realms available '… to the logic of private enterprise and individual initiative' (2013, p. 42). Citizens and consumers are thus free to assume responsibility over their decision making within the constraints of market logics (Joseph 2013, p. 42) and:

> Resilience contributes to this through its stress on heightened self-awareness, reflexivity and responsibility. It encourages the idea of active citizenship, whereby people, rather than relying on the state, take responsibility for their own social and economic well-being.

Davoudi (2012) reaches a similar conclusion. Resilience thinking's self-organisation, she writes, '... becomes highly charged with ideological overtones' (Davoudi 2012, p. 305) when realised in ideology, wherein it becomes self-reliance. She finds that if interpreted as implying the retreat of government responsibilities, that self-reliance matched a neoliberal agenda, and identified it as '... a kind of social Darwinism' (Davoudi 2012, p. 305). Walker and Cooper (2011) make the stronger and more explicit case that the foundations of the resilience concept share a common ideological and historical foundation with what became neoliberalism, notably through the interchange of ideas by their two respective founders, the ecologist Crawford Holling and economist Friedrich Hayek. Given the extensive scholarship on resilience and neoliberalism, the apparent neglect of Walker and Cooper's findings are somewhat surprising.

## Socio-Ecological System (SES) Transformations' Implicit Politics

Identifying the politics of transformative socio-ecological adaptations is riddled with the ambiguity of the SES concept, or rather the range of competing visions (Herrero-Jáuregui et al. 2018), and this may have inhibited political analysis. Further 'identity' problems arise from the multi-dimensional aspects of transformation (such as changes to culture, economics, institutions and technology). And as described above, for many scholars, transformation is a multi-scalar phenomenon, a factor that further complicates analysis of power and politics (Boonstra 2016). There are also competing views as to what the implicit politics of transformation might comprise, such as whether it concerns political processes or political outcomes.

Three broad approaches appear possible for this analysis: Politics that enables transformation (i.e., the normative project), politics of transformation and the transformation of politics itself (such as changes in governance) (Manuel-Navarrete and Pelling 2015). Such approaches would depict transformations as political activities involving political institutions, political actors and political values and producing political outcomes. Generally, scholarship into transformative adaptations, however, does not place politics at the centre of description and analysis—but has it as a either a peripheral, residual or incidental factor; oftentimes it is identified only as an amorphous barrier to successful transformation.

Pelling (2011) adopted a multi-pronged use of three contrasting theories to describe the politics of transformation by applying (Ulrich Beck's) risk society theory, a human security framework and the social contract (see also, Manuel-Navarrete and Pelling 2015). Only the social contract can be fitted into the familiar constellation of political ideologies given its links with (classical) liberalism; it is difficult to establish the other frameworks' political implications. Pelling (2011) offers that transformative change following some natural disasters are indicative of the failure of the extant social contract. It appears that the manifestation of a social contract is most apparent when the arrangements of rights and responsibilities are clear and this is associated with defined jurisdictions and actors with formally assigned roles. Pelling (2011) effectively concedes this point by explaining that applying the social contract can be problematic, especially given the difficulty of defining rights and responsibilities in real world conditions with diverse actors and multiple scales of governance.

In many ways, the aforedescribed political critiques of resilience may be applicable to transformation, given that both share a weakly theorised notion of social systems compared to that expected in the social sciences. One aspect of this critique is rooted in systems theory approaches to social problems, namely its apolitical stance and the inclination to accept extant social relations uncritically. SES models may also be far more adept in locating adaptation failures based in socio-ecological relations, but far less so than for those solely based in social relations (notably those involving power and influence). Just as fundamentally, social systems exhibit features that do not exist in ecological systems so that socio-ecological models are insubstantial in treating social institutions, actors' behaviour and various distinct social functions. As such, while these systems include the material aspects of adaptation, the non-material and social values and tend to escape coverage, such as political (and other) values, the distribution of wealth, spiritual interests, and learning and innovation. Following these points, from an analytical perspective, SES suffer from narrowness in their explanations.

In the following chapter, on political analyses of adaptation, the influence of political ideologies and values is made more explicit.

## References

Abeygunawardena, P., Vyas, Y., Knill, P., Foy, T., Harrold, M., Steele, P., et al. (2009). *Poverty and climate change: Reducing the vulnerability of the poor through adaptation*. Washington, DC: World Bank.

Adams, V. (2012). The other road to serfdom: Recovery by the market and the affect economy in New Orleans. *Public Culture, 24*(1), 185–216.

Adger, W. N., Paavola, J., Huq, S., & Mace, M. J. (Eds.). (2006). *Fairness in adaptation to climate change*. Cambridge: MIT Press.

Adger, W. N., Barnett, J., Chapin, F. S., & Ellemor, H. (2011). This must be the place: Underrepresentation of identity and meaning in climate change decision-making. *Global Environmental Politics, 11*(2), 1–25.

Adger, W. N., Butler, C., & Walker-Springett, K. (2017). Moral reasoning in adaptation to climate change. *Environmental Politics, 26*(3), 371–390.

Agrawal, A., & Perrin, N. (2008). *Climate adaptation, local institutions, and rural livelihoods* (IFRI Working paper W081-6). University of Michigan.

Aguilar, L. (2009). *Training manual on gender and climate change*. UNDP: IUCN.

Andrade Pérez, A., Herrera Fernández, B., & Roberto Cazzolla Gatti, R. (Eds.). (2010). *Building resilience to climate change: Ecosystem-based adaptation and lessons from the field*. Gland, Switzerland: IUCN.

Araral, A. (2014). Ostrom, Hardin and the commons: A critical appreciation and a revisionist view. *Environmental Science & Policy, 36*, 11–23.

Arora-Jonsson, S. (2011). Virtue and vulnerability: Discourses on women, gender and climate change. *Global Environmental Change, 21*, 744–751.

Baja, K., & Granberg, M. (2018). From engagement to empowerment: Climate change and resilience in the city of Baltimore. In S. Moloney, H. Fünfgeld, & M. Granberg (Eds.), *Local action on climate change: Opportunities and constraints* (pp. 126–145). London and New York: Routledge.

Beetham, D. (2013). *The legitimation of power* (2nd ed.). Basingstoke: Palgrave Macmillan.

Beetham, D. (2018). Democracy and social justice. In G. Craig (Ed.), *Handbook on global justice* (pp. 66–78). Cheltenham: Edward Elgar.

Benhabib, S. (1994). Deliberative rationality and models of democratic legitimacy. *Constellations, 1*(1), 46–52.

Berkes, F. (2008). Commons in a multi-level world. *International Journal of the Commons, 2*(1), 1–6.
Berkes, F. (2012). Implementing ecosystem-based management: Evolution or revolution? *Fish and Fisheries, 13*(4), 465–476.
Boonstra, W. J. (2016). Conceptualizing power to study social-ecological interactions. *Ecology and Society, 21*(1), 21. https://doi.org/10.5751/ES-07966-210121.
Ciampi, C. M., Gell, F., Lasap, L., & Turvill, E. (2011). *Gender and disaster risk reduction: A training pack*. Oxford: Oxfam.
Colls, A., Ash, N., & Ikkala, N. (2009). *Ecosystem-based adaptation: A natural response to climate change*. Gland, Switzerland: IUCN.
Davoudi, S. (2012). Resilience: A bridging concept or a dead end? *Planning Theory & Practice, 13*(2), 299–307.
Demetriades, J., & Esplen, E. (2008). The gender dimensions of poverty and climate change adaptation. *IDS Bulletin, 39*(4), 24–31.
Denton, F. (2002). Climate change vulnerability, impacts, and adaptation: Why does it matter? In R. Masika (Ed.), *Gender, development, and climate change* (pp. 10–20). Oxford: Oxfam.
Dessai, S., Hulme, M., Lempert, R., & Pielke, R. (2009). Climate predictions: A limit to adaptation? In W. N. Adger, I. Lorenzoni, & K. O'Brien (Eds.), *Adapting to climate change: Thresholds, values, governance* (pp. 64–78). Cambridge: Cambridge University Press.
Dietz, T., Ostrom, E., & Stern, P. C. (2003). The struggle to govern the commons. *Science, 302*, 1907–1912.
Dodman, D., & Mitlin, D. (2013). Challenges for community-based adaptation: Discovering the potential for transformation. *Journal of International Development, 25*, 640–659.
Dreier, P. (2006). Katrina and power in America. *Urban Affairs Review, 41*(4), 528–549.
Dumaru, P. (2010). Community-based adaptation: Enhancing community adaptive capacity in Druadrua Island, Fiji. *WIREs Climate Change, 1*, 751–763.
Ebi, K., & Semenza, J. (2008). Adaptations and solutions: Community-based adaptations to the health impacts of climate change. *American Journal of Preventative Medicine, 35*(5), 501–507.
Enarson, E., Fothergill, A., & Peek, L. (2006). Gender and disaster: Foundations and directions. In H. Rodriguez, E. L. Quarantelli, & R. R. Dynes (Eds.), *Handbook of disaster research* (pp. 130–146). New York: Springer.
Ensor, J. E., Park, S. E., Atwood, S. J., Kaminski, A. M., & Johnson, J. E. (2018). Can community-based adaptation increase resilience? *Climate and Development, 10*(2), 134–151.

Fazey, I., Wise, R. M., Lyon, C., Câmpeanu, C., Moug, P., & Davies, T. E. (2016). Past and future adaptation pathways. *Climate and Development, 8*(1), 26–44.

Forsyth, T., & Johnson, C. (2014). Elinor Ostrom's legacy: Governing the commons, and the rational choice controversy. *Development and Change, 45*(5), 1093–1110.

Foster, S., Leichenko, R., Nguyen, K. H., Blake, R., Kunreuther, H., Madajewicz, M., et al. (2019). New York City panel on climate change report chapter 6: Community-based assessments of adaptation and equity. *Annals of the New York Academy of Sciences, 1439,* 126–173.

Fraser, N. (2001). Recognition without ethics. *Theory, Culture & Society, 18*(2–3), 21–42.

Gaillard, J. C., Sanz, K., Balgos, B. C., Dalisay, S. N. M., Gorman-Murray, A., Smith, F., et al. (2017). Beyond men and women: A critical perspective on gender and disaster. *Disasters, 41*(3), 429–447.

Granberg, M., & Glover, L. (2014). Adaptation and maladaptation in Australian national climate change policy. *Journal of Environmental Policy and Planning, 16*(2), 147–159.

Grove, K. (2016). Catastrophic insurance and the biopolitics of climate change adaptation. In S. O'Lear & S. Dalby (Eds.), *Reframing climate change: Constructing ecological geopolitics* (pp. 171–187). London and New York: Routledge.

Hale, B., Hermans, A. P., & Lee, A. (2013). Climate adaptation, moral reparation, and the baseline problem. In S. C. Moser & M. T. Boykoff (Eds.), *Successful adaptation to climate change: Linking science and policy in a rapidly changing world* (pp. 67–80). London: Routledge.

Harcourt, W. (Ed.). (1994). *Feminist perspectives on sustainable development.* London and Atlantic Highlands, NJ: Zed Books.

Hegger, D., & Dieperink, C. (2016). Toward successful joint knowledge production for climate change adaptation: Lessons from six regional projects in the Netherlands. *Ecology and Society, 19*(2), 34.

Heller, N. E., & Zavaleta, E. S. (2009). Biodiversity management in the face of climate change: A review of 22 years of recommendations. *Biological Conservation, 142*(1), 14–32.

Herrero-Jáuregui, C., Arnaiz-Schmitz, C., Fernanda Reyes, M., Telesnicki, M., Agramonte, I., Easdale, M. H., et al. (2018). What do we talk about when we talk about social-ecological systems? A literature review. *Sustainability, 10,* 2950. https://doi.org/10.3390/su10082950.

Holling, C. S. (1973). Resilience and stability of ecological systems. *Annual Review of Ecology and Systematics, 4,* 2–23.

Huq, S., Reid, H., & Murray, L. A. (2006). *Climate change and development links.* London: International Institute for Environment and Development.

IPCC. (2001). *Climate change 2001: Impacts, adaptation and vulnerability*. Cambridge: Cambridge University Press.
IPCC. (2007). *Climate change 2007: Impacts, adaptation and vulnerability*. Cambridge: Cambridge University Press.
IPCC. (2014). *Climate change 2014: Impacts, adaptation, and vulnerability*. Cambridge: Cambridge University Press.
Jagers, S., & Duus-Otterström, G. (2008). Dual climate change responsibility: On moral divergences between mitigation and adaptation. *Environmental Politics, 17*(4), 576–591.
Jagers, S., Duus-Otterström, G., & Stripple, J. (2009). Fair and feasible climate change adaptation. *Statsvetenskaplig Tidskrift, 111*(1), 53–58.
Jerneck, A. (2018). What about gender in climate change? Twelve feminist lessons from development. *Sustainability, 10,* 627.
Joseph, J. (2013). Resilience as embedded neoliberalism: A governmentality approach. *Resilience, 1*(1), 38–52.
Kelman, I. (2020). *Disaster by choice: How our actions turn natural hazards into catastrophes*. Oxford: Oxford University Press.
Kinnvall, C., & Rydström, H. (Eds.). (2019). *Climate hazards, disasters, and gender ramifications*. New York: Routledge.
Kirkby, P., Williams, C., & Huq, S. (2017). Community-based adaptation (CBA): Adding conceptual clarity to the approach, and establishing its principles and challenges. *Climate and Development, 10*(7), 577–589.
Lambrou, Y., & Piana, G. (2005). *Gender: The missing component in the response to climate change*. Rome: FAO.
Lopez-Carresi, A., Fordham, M., Wisner, B., Kelman, I., & Gaillard, J. C. (Eds.). (2013). *Disaster management: International lessons in risk reduction, response and recovery*. London: Routledge.
Lukes, S. (2005). *Power: A radical view* (2nd ed.). London: Red Globe Press.
Mansuri, G., & Rao, V. (2012). *Localizing development: Does participation work?* Washington, DC: The World Bank.
Manuel-Navarrete, D., & Pelling, M. (2015). Subjectivity and the politics of transformation in response to development and environmental change. *Global Environmental Change, 35,* 558–569.
Mawdsley, J. R., O'Malley, R., & Ojima, D. S. (2009). A review of climate-change adaptation strategies for wildlife management and biodiversity conservation. *Conservation Biology, 23*(5), 1080–1089.
McGray, H., Hammill, A., & Bradley, R. (2007). *Weathering the storm: Options for framing adaptation and development*. Washington, DC: World Resources Institute.
Mearns, R., & Norton, A. (2010). Equity and vulnerability in a warming world: Introduction and overview. In R. Mearns & A. Norton (Eds.), *Social*

*dimensions of climate change: Equity and vulnerability in a warming world* (pp. 1–44). Washington, DC: World Bank.

Mikulewicz, M. (2018). Politicizing vulnerability and adaptation: On the need to democratize local responses to climate impacts in developing countries. *Climate and Development, 10*(1), 18–34.

Moloney, S., Fünfgeld, H., & Granberg, M. (Eds.). (2018a). *Local action on climate change: Opportunities and constraints*. London and New York: Routledge.

Moloney, S., Fünfgeld, H., & Granberg, M. (2018b). Climate change responses the global to local scale: An overview. In S. Moloney, H. Fünfgeld, & M. Granberg (Eds.), *Local action on climate change: Opportunities and constraints* (pp. 1–16). London and New York: Routledge.

Munang, R., Thiaw, I., Alverson, K., Mumba, M., Liu, J., & Rivington, M. (2013). Climate change and ecosystem-based adaptation: A new pragmatic approach to buffering climate change impacts. *Current Opinion in Environmental Sustainability, 5*(1), 67–71.

Nellemann, C., Verma, R., & Hislop, L. (Eds.). (2011). *Women at the frontline of climate change: Gender risks and hopes*. GRID-Arendal: United Nations Environment Programme.

Nelson, V., Meadows, K., Cannon, T., Morton, J., & Martin, A. (2002). Uncertain predictions, invisible impacts, and the need to mainstream gender in climate change adaptations. *Gender and Development, 10*(2), 51–59.

Nightingale, A. J. (2017). Power and politics in climate change adaptation: Struggles over authority and recognition in the context of political instability. *Geoforum, 84*, 11–20.

Olsson, D. (2018). *Conditions of 'sustainability': The case of climate change adaptation in Sweden*. Karlstad: Karlstad University Press.

Ostrom, E. (1990). *Governing the commons: The evolution of institutions for collective action*. Cambridge: Cambridge University Press.

Paavola, J. (2005). Seeking justice: International environmental governance and climate change. *Globalizations, 2*(3), 309–322.

Paavola, J., & Adger, W. N. (2006). Fair adaptation to climate change. *Ecological Economics, 56*(4), 594–609.

Pelling, M. (2011). *Adaptation to climate change: From resilience to transformation*. London: Routledge.

Peterson, G. (2009). Ecological limits of adaptation to climate change. In N. W. Adger, I. Lorenzoni, & K. L. O'Brien (Eds.), *Adapting to climate change: Thresholds, values, governance* (pp. 25–41). Cambridge: Cambridge University Press.

Preet, R., Nilsson, M., Schumann, B., & Evengård, B. (2010). The gender perspective in climate change and global health. *Global Health Action, 3*(1), 5720.

Quigley, W. P. (2008). What Katrina revealed. *Harvard Law and Policy Review, 2*(2), 361–384.
Reeves, D. (2005). *Planning for diversity: Policy and planning in a world of difference*. London: Routledge.
Reid, H. (2016). Ecosystem- and community-based adaptation: Learning from community-based natural resource management. *Climate and Development, 8*(1), 4–9.
Reid, H., & Huq, S. (2014). Mainstreaming community-based adaptation into national and local planning. *Climate and Development, 6*(4), 291–292.
Reid, H., Alam, M., Berger, R., Cannon, T., Huq, S., & Milligan, A. (2009). Community-based adaptation to climate change: An overview. In H. Reid, T. Cannon, R. Berger, M. Alam, & A. Milligan (Eds.), *Community-based adaptation to climate change* (pp. 11–33). Participatory Learning 60. London: International Institute for Environment and Development.
Reid, J. (2013). Interrogating the neoliberal biopolitics of the sustainable-development-resilience nexus. *International Political Sociology, 7*, 353–368.
Resurrección, B. P. (2013). Persistent women and environment linkages in climate change and sustainable development agendas. *Women's Studies International Forum, 40*, 33–43.
Ribot, J. C. (1995). The causal structure of vulnerability: Its application to climate change analysis. *GeoJournal, 35*(2), 119–122.
Rickards, L. (2010). *Governing the future under climate change: Contested visions of climate change adaptation*. Melbourne: The Victorian Centre for Climate Change Adaptation Research.
Ruhl, J. B. (2012). The political economy of climate change winners. *Minnesota Law Review, 97*, 207–277.
Schlosberg, D. (2011). Climate justice and capabilities: A framework for adaptation policy. *Ethics & International Affairs, 26*(4), 445–461.
Sen, A. (1999). *The development of freedom*. Oxford: Oxford University Press.
Skutsch, M. (2002). Protocols, treaties, and action: The 'climate change process' viewed through gender spectacles. *Gender and Development, 10*(2), 30–39.
Smith, J. B., Vogel, J. M., & Cromwell, J. E. (2009). An architecture for government action on adaptation to climate change. *An Editorial Comment. Climatic Change, 95*(1–2), 53–61.
Sovacool, B. K., & Linnér, B.-O. (2016). *The political economy of climate change adaptation*. Basingstoke: Palgrave Macmillan.
Sowers, J., Vengosh, A., & Weinthal, E. (2011). Climate change, water resources, and the politics of adaptation in the Middle East and North Africa. *Climatic Change, 104*, 599–627.
Stern, P. (2011). Design principles for global commons: Natural resources and emerging technologies. *International Journal of the Commons, 5*(2), 213–232.

Tanner, T., & Allouche, J. (2011). Towards a new political economy of climate change and development. *IDS Bulletin, 42*(3), 1–14.

Terry, G. (2009). No climate justice without gender justice: An overview of the issues. *Gender & Development, 17*(1), 5–18.

Thomas, D. S. G., & Twyman, C. (2005). Equity and justice in climate change adaptation amongst natural-resource-dependent societies. *Global Environmental Change, 15*(2), 115–124.

Thompkins, E. L., & Adger, W. N. (2004). Does adaptive management of natural resources enhance resilience to climate change? *Ecology and Society, 9*(2), 10.

Tol, R. S. J., Downing, T. E., Kuik, O. J., & Smith, J. B. (2004). Distributional aspects of climate change impacts. *Global Environmental Change, 4*(3), 259–272.

Tschakert, P., & Machado, M. (2012). Gender justice and rights in climate change adaptation: Opportunities and pitfalls. *Ethics and Social Welfare, 6*(3), 275–289.

UNDP. (2010). *Community-based adaptation to climate change*. New York: UNDP.

UNESCO. (2017). *Declaration of ethical principles in relation to climate change*. Paris: UNESCO.

UNFCCC. (1992). *United Nations framework convention on climate change*. New York: UN.

UNISDR. (2009). *UNISDR terminology on disaster risk reduction*. Geneva: UNISDR.

Vale, L. J. (2014). The politics of resilient cities: Whose resilience and whose city? *Building Research & Information, 42*(2), 191–201.

Vignola, R., Locatelli, B., Martinez, C., & Imbach, P. (2009). Ecosystem-based adaptation to climate change: What role for policy-makers, society and scientists? *Mitigation and Adaptation of Strategies for Global Change, 14*(8), 691–696.

Villagrasa, D. (2002). Kyoto Protocol negotiations: Reflections on the role of women. In R. Masika (Ed.), *Gender, development and climate change* (pp. 40–44). Oxford: Oxfam.

Vink, M. J., Dewulf, A., & Termeer, C. (2013). The role of knowledge and power in climate change adaptation governance: A systematic literature review. *Ecology and Society, 18*(4), 46.

Waldron, J. (1992). Minority cultures and the cosmopolitan alternative. *University of Michigan Journal of Law Reform, 25*(3–4), 751–793.

Walker, J., & Cooper, M. (2011). Genealogies of resilience: From systems ecology to the political economy of crisis adaptation. *Security Dialogue, 42*, 143–160.

Wall, D. (2017). *Elinor Ostrom's rules for radicals: Cooperative alternatives beyond markets and states*. London: Pluto Press.

WCED. (1987). *Our common future*. New York: The United Nations.
West, J. M, Julius, S. H., Kareiva, P., Enquist, C., Lawler, J. J., Petersen, B., et al. (2009). U.S. natural resources and climate change: Concepts and approaches for adaptation management. *Environmental Management*, *44*(6), 1001–1021.
Wester, M., & Doma Lama, P. (2019). Women as agents of change? Reflections on women in climate adaptation and mitigation in the Global North and the Global South. In C. Kinnvall & H. Rydström (Eds.), *Climate hazards, disasters, and gender ramifications* (pp. 67–85). New York: Routledge.
Wilkinson, C. (2012). Urban resilience: What does it mean in planning practice? *Planning Theory & Practice*, *13*(2), 319–324.

CHAPTER 6

# Political Analyses of Adaptation

**Abstract** Many political analyses of adaptation politics have been undertaken, ranging across an array of issues, locations and subjects and using a variety of political approaches, frameworks and concepts. In this chapter, four prominent and contrasting approaches are examined as these encompass much of this political scholarship: Institutional reform of decision-making, political economy, environmental justice and political ecology. Although each approach is distinct, there are some aspects shared between the approaches and these are described. A range of features of each approach are described, canvassing such aspects as ideology, political focus, key concerns, political realms of interest and normative goals. This chapter draws on the political context and political themes described in the preceding chapters.

**Keywords** Environmental justice · Ideology · Institutional reform · Political ecology · Political economy · Values

Following the aforementioned expansion in adaptation scholarship and the expression of political aspects, a number of studies have explicitly examined adaptation policy as expressing political ideologies. Four prominent approaches are identified:



L. Glover and M. Granberg,
*The Politics of Adapting to Climate Change*,
https://doi.org/10.1007/978-3-030-46205-5_6

- Institutional reform of decision making
- Political economy
- Environmental justice, and
- Political ecology (see, Table 6.1).

This listing is not comprehensive; further, delimiting boundaries between these approaches are fluid and indeterminate in some ways. While some studies may ideologically self-identify, others do not explicitly endorse a specific ideology. Notwithstanding these caveats, each aforementioned category can express distinctive political aspects such as ideology, political focus, key concerns, identity of key political agents, political realms of interest, normative goals, critique of existing practices and identified implications.

## Institutional Reform

Social studies of science scholars have considered issues of creating, identifying, disseminating and applying climate change science at an early stage of the contemporary climate change debate. A number of studies raised issues of the constitution, identity, role and intersecting influence of government and science institutional bodies (e.g., Shackley and Wynne 1995; Raynor and Malone 1998; Demeritt 2001). As such, they were reflecting the 'new institutionalism' in political science that came to prominence in the mid-1990s (cf., March and Olson 1984; Hall and Taylor 1996). New institutionalism was in many respects a rejection of the prevailing behavioralist/utilitarian and socially determinist/Marxist political theories (see, Immergut 1998). Institutions are rules and organised practices guided by norms and values that structure meaning and frame action in specific situations (March and Olson 1984, 2009). A central aspect is that institutions create stability over time in ways that make contemporary decision making framed by, and connected to, decisions made in the past (i.e., path dependency) (Peters 2019). Hence, a new institutional approach can be highly relevant when studying the political aspects of collective action, such as climate change adaptation. It has not been until more recently that politics and public administration scholars have considered institutional matters as they pertain explicitly to adaptation research, policy and planning (Huitema et al. 2016). For scholars

**Table 6.1** Political perspectives on adaptation policy and practices

| | *Concerns* | *Perception of politics* | *Agents* | *Political realm* | *Perception of Nature* | *Adaptation in political context* |
|---|---|---|---|---|---|---|
| Institutional reform | Potential to balance expert knowledge and politics | An influence | Experts<br>Bureaucrats<br>Politicians | Public sector decision-making | A source of societal risk | Adaptation responses reflect political values and interests |
| Political economy | Potential for social and economic injustice | An outcome | International/National agencies<br>Communities<br>Individuals | Projects/actions through international/national agencies | A medium/facilitator of injustice | Adaptation policies and practices have produced social and economic injustice |
| Environmental justice | Potential for environmental injustice | Extended beyond the social realm | Governments<br>Corporations<br>NGOs | International/national policy; broader society; Nature | A victim of injustice | Adaptation policies and practices have produced social and economic injustice |
| Political ecology | Potential to reduce society's domination over nature | As domination | Corporations (capital)<br>Government<br>Nature | The nature society interface | Equal | Adaptation has to adjust to nature |

working with this theme, political concerns are constrained to an institutional setting in studies, usually dealing with policy formulation and implementation.

Institutions, in this context, are understood in a formal context as being those entities for delivering governance through formulating, implementing and monitoring of public policies (and related instruments). Prominent political interests are the effects of political power and political influence, especially differential access to decision making and the exclusion of those of socio-economic disadvantage. Decision making for public policy processes routinely employ government agencies, key corporate stakeholders, technical and scientific experts and 'peak' NGOs representing corporate and environmental group interests. This is discourse focused on knowledge creation and its legitimation.

Two themes predominate in the studies of the governing institutions for adaptation:

- Political values expressed in decision making, and
- Identity and authority of decision makers.

Both themes give rise to two related concerns, one being somewhat the inverse of the other. Firstly, proponents of democracy consider that technological expertise will trump the representatives of the public in decision making on matters requiring specialist knowledge and training (i.e., technocracy). Secondly, that science, especially in its application to matters involving public policy, is corrupted by political influences within the policy-making sphere (or by scientists themselves) and made to serve vested interests (i.e., politicisation).

Tangney (2017) puts adaptation policy into the political realm by seeing how adaptation policy—and the evidence it is based on—is an outcome of political factors. He argues that 'technological' evidence used in climate change decision making is, in fact, politicised in ways that ensure the utilised evidence is coherent with prevailing political and economic ideals (Tangney 2017). He sets this argument up as a classic dichotomy between technocratic expertise and democratic ideals and its application to the problems of decision making and policy formulation for sustainable development. This, again, highlights the importance of path dependency as phenomenon that gives a context to decision making that frames the scope of decisions (Peters 2019). For example, a strong focus on technocratic solutions leads decision making down certain pathways, or a strong focus on democratic ideals opens up other pathways.

Tangney is interested in how public authorities make decisions over adaptation as contests between technical and political authorities. Regarding the desired qualities of the requisite knowledge systems (Tangney 2017, p. 2):

> Some have argued that these attributes should be pursued during the development of policy knowledge in a way that ensures an appropriate balance between the need for technically adequate objective expert knowledge (such that it exists) on the one hand, and evidence that meets the practical and normative requirements of decision-makers on the other.

Politics enters the picture as it plays a role in legitimising knowledge. Tangney offers two pathways for the politicisation of climate change adaptation knowledge. One pathway is the politicisation-by-process resulting from the value judgements and subjective decisions of experts in providing evidence to policymakers. Critically, Tangney is referring to inputs that exceed those routinely used in science practices and he calls these 'normative' inputs. This contrasts with the second pathway that is characterised politicisation-by-agency, where (expert and non-expert) officials intervene in the results or communications of adaptation knowledge motivated by political considerations.

Studies of such institutional workings represent a fraction of the overall scholarship on adaptation. Investigations of the mechanics of public policy processes and their associated knowledge production and use are few, especially in light of the number of jurisdictions currently producing adaptation policies. To the high-profile political issues in adaptation, studies of the public administration of adaptation responses and of organisations (such as accountability, financing, leadership and management) the machinery of governments can add a neglected and influential aspect of adaptation politics. Numerous studies have described the privileging of scientific knowledge in public policy, but typically these offer 'outside' perspectives, as opposed to intensive studies of the workings of governance institutions (such as the activities and influence of key stakeholders, the role of funding, inter-agency influences and the like). Understanding of how governments have developed institutions (as distinct from developing policies and programmes) for adaptation governance is also a somewhat neglected topic in scholarship; exceptions include Edelenbos (2012) study of Dutch institutions and Termeer et al. (2012).

Public participation (or, alternatively, 'inclusive', 'stakeholder-engaged' or 'bottom-up' decision making processes) has been offered as a means to break up the path dependency, or to mitigate or alleviate, the aforementioned technocratic, specialist, centrist and elitist characteristics of adaptation policy formulation and planning processes (Reid et al. 2009; Terry 2009; Pelling 2011; Tanner and Allouche 2011; Agrawal et al. 2012; Reid and Huq 2014; Sovacool and Linnér 2015; Kirkby et al. 2017). Making institutions take wider or particular community interests into account can enable these institutions—as forms of democratic process—to counteract the influence of the aforementioned special interests.

Indeed, a wide array of national and international agencies and NGOs have consistently called for such participation (e.g., IPCC 2001, 2014), to the extent that can be regarded as normal rhetorical practice (some of which extends into 'post-normal' science). However, it is also widely observed that most participatory practices have fallen considerably short of the ideals and expectations, due to pragmatic limitations and various barriers and variations in participatory models and practices (Few et al. 2006; Mansuri and Rao 2012; Ensor et al. 2018; Kirkby et al. 2017).

These shortfalls in public participation practices have overt political dimensions, reflecting some of the limitations of community-based environmental planning more generally (e.g., Lane and McDonald 2005). Ensuring broad consultation can fail in practice, with special-interest groups and advantaged participants exerting proportionally greater influence and the effect of authorities pre-selecting the themes, issues and options for public consideration (Adger et al. 2005b). Community-based adaptation offers an approach that can potentially be more inclusive, but not necessarily so (e.g., Few et al. 2006; Reid et al. 2009; Kirkby et al. 2017). Nagoda and Nightingale (2017) found that participatory practices in a Nepalese adaptation program were reinforcing—rather than alleviating—existing vulnerabilities. Consultation programmes can be 'shallow', with stakeholder engagement providing the appearance of participation with relatively little substance, thereby reinforcing existing power hierarchies. Reducing the scale of decision making spheres to the local can merely reduce the scale of the political contests over marginalisation, resource access, risk and vulnerability rather than obviating them (i.e., it produces micro-political issues). Other less overtly political issues in public participation in adaptation policies and programs include public apathy and cynicism towards involvement, problems with large-scale programmes and communication gaps.

## Political Economy

Political economy has its foundations in classical economic theory as developed by Adam Smith, David Ricardo, Thomas Malthus and others (Stilwell 2002). Broadened in approach in contemporary times, political economy includes diverse political perspectives, such as neo-Marxism and public choice (free market economics) (i.e., it connotes a theme, interest or approach, not a specific political ideology). Accordingly, many definitions are available, but essentially political economy is concerned with the mutual influence of economics and politics that produce political and economic outcomes, and the ongoing social implications (Stilwell 2002). Common themes are:

- Gaining and expressing of political power and economic influence
- Distribution of power and wealth in society
- Political explanations of resource allocation and possible redistribution
- Economic explanations of political behaviour and structures (including ideas and interests), and
- Roles and activities of state institutions and the activities of corporations (especially the role of political power and influence).

In adaptation politics, studies evoking political economy typically express progressive social goals, normative intentions and a critique of social and economic inequity. Such an orientation for political economy places it within the classical ideology of socialism, as this dictates its core interests, informing a particular critique of society, view of social change and vision for the future (cf., Walker 1994; Hay 1999). Adaptation policy at sub-national, national and supra-national scales clearly evokes many of these fundamental themes and concerns of political economy (noting that there is now widespread activity at the local level) (Moloney et al. 2018). Political economy gives expression to socio-economic factors in three broad sequential phases of adaptation: Defining vulnerability to climate change as a socio-economic phenomenon, formulating and delivering adaptation measures as a socio-economic process and post-adaptation distribution of gains and losses; these are reviewed in turn.

Firstly, socio-economic circumstances determine vulnerability, with risk being a function of such factors as capital (and social capital), access to political power, government policies and knowledge. Accounts of these risks are usually a part of adaptation studies with wider interests. Because the greatest vulnerabilities tend to be concentrated amongst poorer communities in poorer nations, political economy is particularly interested in understanding the socio-economic circumstances of these peoples, notably those inequalities in the economic, political and social realms, concentrations of wealth and power, and the circumstances of disadvantaged groups. For example, individual and community ability to prepare, respond and recover from disasters evokes the political economy variables, including age, class, economic resources, ethnicity, gender, health, land tenure and social status.

Secondly, political economy analyses focus on features of adaptation planning, policy and program formulation, and project delivery. Adaptation policy is built upon a strongly economically-rational foundation and this informs how risks, vulnerabilities and priorities are understood (Fieldman 2011; Granberg and Nyberg 2018). This rationality is expressed in decisions over setting protection priorities and in how this protection should occur (and conversely, what and whom are not protected). Valuable economic assets and activities are accorded high priorities, as these are deemed to be important in their own right and as contributors to local or larger economies (Storbjörk and Hjerpe 2013). Natural resources and ecosystem services with high economic values are also given high priorities. This occurs in several ways, most obviously through measures to protect assets and major investments of public and private capital. Hence, the common use of cost-benefit analysis in vulnerability and adaptation assessment studies.

Thirdly, adaptation measures can be an agent of socio-economic conditions. Adaptation achieves its agency by influencing the pre-existing socio-economic conditions, by redistributing material and other resources under climate change impacts through adaptation measures (i.e., as a mediator of influence) and directly through the additional costs and benefits arising from the adaptation measures (i.e., the distribution of progressive and regressive effects). Along with its material influences, political economy also investigates the influence of adaptation on political and economic power and influence. Sovacool et al. (2015, p. 617) articulated this outcome as to '… aggravate political, socio-economic, or cultural, inequalities and the disempowerment of disadvantaged groups' leading to '… concentrating the wealth within a community'. Adger et al. (2005a) set out criteria for evaluating adaptation measures and in dealing with (economic) efficiency as a criterion. Yardsticks for adaptation measures require knowledge of the distribution of costs and benefits, of non-market costs and benefits, and of timing. Three issues are identified by Adger et al. (2005a):

- Fluidity of the boundary between public and private goods (using these terms in their economic sense)
- Estimates of non-market benefits and costs (notably of social and environmental phenomena) tend to be underestimated, and
- Timing of adaptation actions can greatly influence the efficiency of outcomes (prominently, short- and long-term differences in planning horizons, climate change impacts and economic evaluations).

Sovacool and Linnér (2015) in *The Political Economy of Climate Change Adaptation* provide the most systematic and broad ranging analysis from this perspective, beginning with their categorisation of the types of political science adaptation studies:

- Corruption (e.g., Sweeney et al. 2011)
- Maladaptation (e.g., Barnett 2010)
- 'Winners and losers' (e.g., Ruhl 2012)
- Responsibility for GHG emissions vs. vulnerability to climate change impacts dichotomies (e.g., Althur et al. 2016)
- Sustainable and unsustainable adaptation (e.g., Eriksen et al. 2011, 2015)
- Trade-offs between emissions mitigation and adaptation (e.g., Wilbanks and Sathaye 2007), and
- Discourse studies (e.g., Adger et al. 2001; Bankoff 2001).

Sovacool and Linnér (2015), Sovacool et al. (2015) and Sovacool (2018) offer four processes/outcomes to describe the how inequity is generated in contrasting socio-ecological realms in wide-ranging case studies of adaptation:

- Enclosure (in the economic realm): Capture and transfer of resources or authority, privatisation
- Exclusion (in the political realm): Excluding/restricting engagement of stakeholders in decision making
- Encroachment (in the ecological realm): Environmental harm, and
- Entrenchment (in the social realm): Disempowering women and minorities, worsening inequality.

Lessons drawn by the authors are consistent with political economy findings elsewhere, beginning with the inexorable entanglement of adaptation processes with the allocation and distribution of resources and political power; adaptation decisions have political consequences. Furthermore, the effects of adaptation decisions can resonate much more widely than the immediate site of their application, affecting other stakeholders, communities, ecosystems and the future.

In calling for a 'new' political economy of climate change, Tanner and Allouche (2011) characterise such an approach in the following terms that include: Policy processes are complex and informed by ideology, actors and power relations; Dominant scale involves translating high-level activity to sub-national and local levels; Research involves the social construction of science; Scarcity and poverty is viewed through political processes that mediate competing claims and decision making is driven by ideology, incentives and power relations. By these measures, it is not unreasonable to consider that political economy has made considerable progress in adaptation research in recent years. A part of this growth stems from the influence of risk and disaster studies.

Political economy does, however, have a number of blind spots. In the manner of its founding ideas, political economy views society rationally through a lens of materiality and tends to confine its conceptions of the ideal to that of material equity and access to power, with no place for the cultural or non-material aspirations. Nature too, is typically (but not always) without an advocate in political economy; it is a source of ecosystem goods and services. Contrasting with institutional studies, political economy pays little attention to decision making means or participation, focusing usually on the influence of power and wealth as socio-economic pre-conditions to adaptation and social justice outcomes of adaptations. It is also less concerned with the means of adaptation, such as the role of science and technology. To some extent, these specific concerns are taken up in environmental justice approaches. Questions involving the values of equity and participatory decision making are strongly covered, and a number of studies examine how political processes deal with such disadvantaged groups, notably those of higher vulnerability, notably the poor, indigenous peoples and women. Commons management tends to have a lower profile in this discourse and is more likely to be found in environmental justice and political ecology discourses.

## Environmental Justice

Environmental justice encompasses a range of environmental politics views and beliefs, overlapping significantly with the concerns and outlooks of political economy. Oftentimes, the notions of justice are implicit in political economy, whereas environmental justice usually entails inquiries into the construction, framing and rationalising of justice; it also more open to recognising non-material values (Schlosberg 2013; Schlosberg and Collins 2014; Shi et al. 2016). Needless to add, the environmental justice discourse is commonly linked to ethics and environmental philosophy; in cases, adaptation is posed an as ethical challenge (Ciplet and Roberts 2017; Grasso 2010).

Environmental justice approaches to adaptation are within the general environmental justice movement, being concerned with justice in pre-existing conditions, within procedural processes and as an outcome. Schlosberg (2007, p. 12) considers that justice, in theory and practice, should comprise '... a balance of numerous interlinked elements of distribution, recognition, participation, and capability.' Much of what is covered under environmental justice can be tied to the concept of 'climate justice'. (Schlosberg and Collins (2014) explain the roots of the climate justice has three 'conceptualisations': academe, elite NGOs and grassroot movements.) Most attention in this discourse is directed towards socio-economic inequalities, but not exclusively, as participating in decision making, maintaining cultural integrity and ensuring key social functions are also key interests. Political topics prominent in political economy taken up in environmental justice include those of equity, decision making participation and access, gender, indigenous peoples, and power and influence. Socio-economic differentiation in risk exposure, impacts of neoliberalism, priority-setting in adaptation policies and vulnerability are topics of particular interest.

Climate justice movements' interest in adaptation begins, in a sense, with recognising the prevailing injustices in extant social settings at all scales and across societies (from the individuals, groups and communities to geographic areas to political jurisdictions). Onto these socio-economic circumstances can be mapped the injustices in the distribution and intensity of vulnerabilities to climate change impacts; i.e., not only is vulnerability socially constructed, but these outcomes can be subjectively understood as expressions of injustice and unfairness. Adaptation activities, whether in anticipation or in response to the experiences of climate change impacts, not only can be interpreted in terms of their implications for justice, but also as an overlay on the pre-existing injustices.

Many studies evoking environmental justice do so without extending justice for the environment for its own sake but consider the natural world as having only instrumental or utilitarian values, thereby expressing a narrow version of environmental justice. In part, this reflects the history of the environmental justice concept and social movement, where it began with a concern over inequities in the social exposure by such factors as race and class, to environmental risks and primarily a particular expression of social injustice, notably in urban contexts and from local industrial hazards (Schlosberg 2013). These concerns expanded through a broader conception of the 'environment', consideration of the production of injustice and a wider understanding of justice (Schlosberg 2013). One dimension of environmental justice is the needs of the environment, i.e., recognising intrinsic values of the natural world. Part of transformative adaptation agenda now involves ideas of ecological restoration, with complex ethical issues over social responsibilities for Nature, notions of flourishing and recognising humanity as a major evolutionary force in forging 'novel' natural systems (Harris et al. 2006).

## Political Ecology

Even by the semantically challenging standards of the environmental, sustainability and development discourses, 'political ecology' is a tricky concept. Most commentators agree that political ecology is without a singular, encompassing theory (Paulson et al. 2003; Robbins 2012). Scholars (and environmental activists) with quite varying approaches self-identify in the category; needless to add, it is a widely interpreted concept, as its use in adaptation politics also reveals. Elements of political economy of the environment, environmental justice, SES theory and human ecology can be found in the scholarship identifying with political ecology. For some scholars and authors, political ecology is now an umbrella term describing inquiries into environmental politics, e.g., Sovacool applies 'political economy' and 'political ecology' interchangeably (cf., Sovacool 2018; Sovacool and Linnér 2015; Sovacool et al. 2015).

In some respects, political ecology was a corrective to human ecology approaches that sought to explain existing or changing social relationships with the environment without reference to politics. Its intellectual roots are diverse, but political ecology arose from cultural ecology, critical development research, disaster studies, environmental history, peasant studies and postcolonial theory (see, Robbins 2012; Watts 2000). As a field of inquiry, political ecology encompasses specific notions of Nature-society relations drawn from ecological principles and processes (e.g., Blaikie and Brookfield 1987) to the broad field examining Nature-society-economy relationships in which the identity of the natural is a more open, dynamic and subjective matter (Walker 2005). Its core interests are the political influences in '... environmental access, management, and transformation' (Robbins 2012, p. 3). Watts (2000) approached the field using two themes, that of knowledge, power and practice, and that of politics, justice and governance. Political ecology informed the critical response to the scientific-rational orientation of IPCC's view of vulnerability as constituting the triad of adaptive capacity, sensitivity to impacts and level of risk exposure. O'Brien et al. (2007) differentiated between the 'scientific framing' and 'human security framing' of vulnerability.

In adaptation politics, political ecology is treated as a field of discourse, but without a single approach or unitary set of principles. Studies in this field range across an array of scales (from international to local and 'micro' issues), topics (e.g., politics, economics and ecology), subjects (e.g., sea level rise, agriculture and flooding) and themes (e.g., sustainability, corruption and equity). Taylor's (2014) *The Political Ecology of Climate Change Adaptation* critiques the apolitical perspective common to much adaptation scholarship, arguing that this restricts political inquiry and functions conservatively by taking adaptation as the 'natural' response to ecological problems. Further, adaptation policies and plans are usually premised on a dualistic simplification of the culture-climate relationships. Taylor writes that this produces a view of climate as an external disturbance acting upon a coherent society, thereby obscuring the role of socio-ecological processes (such as capital accumulation, technology change and political contestation) that produce the lived environment.

A product of a specific governmentality, Taylor argues this construct is designed to suit the institutional needs for controlling the procedures of social change, thereby circumventing issues of power hierarchies, vested interests and the like.

Consequently, an expression of political power is the authority of policy-formulating institutions to determine the criteria by which measures are deemed to be reasonable, affordable and practical. These determinations set the boundaries for the range and performance of adaptation measures within the sphere of institutional politics, thereby ensuring that policies will only proffer marginal change and leave the status quo unchallenged. Taylor adopts 'governmentality' (ala Foucault) and posits the case for a radical recasting of adaptation politics. A couple of other themes contend for additional inclusion in this list, notably participation in decision making and its implications. Political ecology encapsulates, therefore, a wide array of studies.

## References

Adger, W. N., Tor, A. B., Brown, K., & Svarstad, H. (2001). Advancing a political ecology of global environmental discourses. *Development and Change, 32,* 681–715.

Adger, W. N., Arnell, N. W., & Thompkins, E. L. (2005a). Successful adaptation to climate change across scales. *Global Environmental Change, 15,* 77–86.

Adger, W. N., Brown, K., & Tompkins, E. L. (2005b). The political economy of cross-scale networks in resource co-management. *Ecology and Society, 10*(2), 9. http://www.ecologyandsociety.org/vol10/iss2/art9/.

Agrawal, A., Perrin, N., Chhatre, A., Benson, C. S., & Kononen, M. (2012). Climate policy processes, local institutions, and adaptation actions: Mechanisms of translation and influence. *WIREs Climate Change, 3,* 565–579.

Althur, G., Watson, J. E. M., & Fuller, R. A. (2016). Global mismatch between greenhouse gas emissions and the burden of climate change. *Scientific Reports, 6,* 20281.

Bankoff, G. (2001). Rendering the world unsafe: 'Vulnerability' as western discourse. *Disasters, 25*(1), 19–35.

Barnett, J. (2010, May/June). Adapting to climate change: Three key challenges for research and policy—An editorial essay. *WIREs Climate Change, 1,* 314–317.

Beck, U. (2005). *Power in the global age: A new global political economy.* Cambridge: Polity.

Blaikie, P. M., & Brookfield, H. (Eds.). (1987). *Land degradation and society.* London and New York: Methuen.

Castree, N. (2010). Neoliberalism and the biophysical environment. *Environment and Society: Advances in Research, 1,* 5–45.

Ciplet, D., & Roberts, J. T. (2017). Climate change and the transition to neoliberal environmental governance. *Global Environmental Change, 46,* 148–156.

Coffey, B. (2016). Unpacking the politics of natural capital and economic metaphors in environmental policy discourse. *Environmental Politics, 25*(2), 203–222.

Demeritt, D. (2001). The construction of global warming and the politics of science. *Annals of the Association of American Geographers, 91*(2), 307–337.

Edelenbos, J. (2012). Institutional implications of interactive governance: Insights from Dutch practice. *Governance, 18*(1), 111–134.

Ensor, J. E., Park, S. E., Atwood, S. J., Kaminski, A. M., & Johnson, J. E. (2018). Can community-based adaptation increase resilience? *Climate and Development, 10*(2), 134–151.

Eriksen, S., Aldunce, P., Bahinipati, C. S., D'Almeida Martins, R., Molefe, J. I., Nhemachena, C., et al. (2011). When not every response to climate change is a good one: Identifying principles for sustainable adaptation. *Climate and Development, 3*(1), 7–20.

Eriksen, S. H., Nightingale, A. J., & Eakin, H. (2015). Reframing adaptation: The political nature of climate change. *Global Environmental Change, 35*(1), 523–533.

Few, R., Brown, K., & Tompkins, E. L. (2006). *Public participation and climate change adaptation* (Tyndall Centre for Climate Change Research, Working Paper 95).

Fieldman, G. (2011). Neoliberalism, the production of vulnerability and the hobbled state: Systemic barriers to climate adaptation. *Climate and Development, 3*(2), 159–174.

Granberg, M. (2018). Strong local government moving to the market? The case of low carbon futures in the city of Örebro, Sweden. In A. Luque-Ayala, S. Marvin, & H. Bulkeley (Eds.), *Rethinking urban transitions: Politics in the low carbon city* (pp. 129–145). London and New York: Routledge.

Granberg, M., & Glover, L. (2014). Adaptation and maladaptation in Australian national climate change policy. *Journal of Environmental Policy & Planning, 16*(2), 147–159.

Granberg, M., & Nyberg, L. (2018). Climate change adaptation, city competitiveness and urban planning in the city of Karlstad, Sweden. In S. Moloney, H. Fünfgeld, & M. Granberg (Eds.), *Local action on climate change: Opportunities and constraints* (pp. 111–125). London and New York: Routledge.

Grasso, M. (2010). An ethical approach to climate adaptation finance. *Global Environmental Change, 20*(1), 74–81.

Hall, P. A., & Taylor, C. R. (1996). Political science and the three new institutionalisms. *Political Studies, 44*(5), 936–957.

Harris, J. A., Hobbs, R. J., Higgs, E., & Aronson, J. (2006). Ecological restoration and global climate change. *Restoration Ecology, 14*(2), 170–176.

Hay, C. (1999). *The political economy of new labour: Labouring under false pretences?*. Manchester: Manchester University Press.

Huitema, D., Adger, W. N., Berkhout, F., Massey, E., Mazmanian, D., Munaretto, S., et al. (2016). The governance of adaptation: Choices, reasons, and effects. *Ecology and Society, 21*(3), 37.

Immergut, E. M. (1998). The theoretical core of the new institutionalism. *Politics & Society, 26*(1), 5–34.

IPCC. (2001). *Climate change 2001: Impacts, adaptation and vulnerability*. Cambridge: Cambridge University Press.

IPCC. (2012). *Managing the risks of extreme events and disasters to advance climate change adaptation*. Cambridge: Cambridge University Press.

IPCC. (2014). *Climate change 2014: Impacts, adaptation, and vulnerability*. Cambridge: Cambridge University Press.

Kirkby, P., Williams, C., & Huq, S. (2017). Community-based adaptation (CBA): Adding conceptual clarity to the approach, and establishing its principles and challenges. *Climate and Development, 10*(7), 577–589.

Lane, M. B., & McDonald, G. (2005). Community-based environmental planning: Operational dilemmas, planning principles and possible remedies. *Journal of Environmental Planning and Management, 48*(5), 709–731.

Larner, W. (2000). Neo-liberalism: Policy, ideology and governmentality. *Studies in Political Economy, 63*(Autumn), 5–25.

Mansuri, G., & Rao, V. (2012). *Localizing development: Does participation work?* Washington, DC: The World Bank.

March, J. G., & Olson, J. P. (1984). The new institutionalism: Organizational matters in political life. *The American Political Science Review, 78*(3), 734–749.

March, J. G., & Olson, J. P. (2009). Elaborating the "new institutionalism". In R. E. Goodin (Ed.), *The Oxford handbook of political science* (pp. 159–175). Oxford: Oxford University Press.

McEvoy, D., & Fünfgeld, H. (2013). Resilience and climate change adaptation: The importance of framing. *Planning Practice & Research, 28*(3), 280–293.

Mol, A. P. J. (2016). The environmental nation state in decline. *Environmental Politics, 25*(1), 48–68.

Moloney, S., Fünfgeld, H., & Granberg, M. (2018). Climate change responses the global to local scale: An overview. In S. Moloney, H. Fünfgeld, & M. Granberg (Eds.), *Local action on climate change: Opportunities and constraints* (pp. 1–16). London and New York: Routledge.

Nagoda, S., & Nightingale, A. (2017). Participation and power in climate change adaptation policies: Vulnerability in food security programs in Nepal. *World Development, 100,* 85–93.

O'Brien, K., Erikesen, S., Nygaard, L. P., & Schjolden, A. (2007). Why different interpretations of vulnerability matter in climate change discourses. *Climate Policy, 7*(1), 73–88.

Paulson, S., Gezon, L. L., & Watts, M. (2003). Locating the political in political ecology: An introduction. *Human Organization, 62*(3), 205–217.

Pelling, M. (2011). *Adaptation to climate change: From resilience to transformation*. London: Routledge.

Peters, B. G. (2019). *Institutional theory in political science: The new institutionalism* (4th ed.). Cheltenham: Edvard Elgar.

Raynor, S., & Malone, E. L. (Eds.). (1998). *Human choice and climate change: Vol. I: The societal framework*. Columbus: Batelle Press.

Reid, H., & Huq, S. (2014). Mainstreaming community-based adaptation into national and local planning. *Climate and Development, 6*(4), 291–292.

Reid, H., Alam, M., Berger, R., Cannon, T., Huq, S., & Milligan, A. (2009). Community-based adaptation to climate change: An overview. In H. Reid, T. Cannon, R. Berger, M. Alam, & A. Milligan (Eds.), *Community-based adaptation to climate change* (pp. 11–33). Participatory Learning 60. London: International Institute for Environment and Development.

Robbins, P. (2012). *Political ecology: A critical introduction* (2nd ed.). Malden, MA: Wiley-Blackwell.

Ruhl, J. B. (2012). The political economy of climate change winners. *Minnesota Law Review, 97*, 207–277.

Schlosberg, D. (2007). *Defining environmental justice: Theories, movements, and Nature*. Oxford: Oxford University Press.

Schlosberg, D. (2013). Theorising environmental justice: The expanding sphere of a discourse. *Environmental Politics, 22*(1), 37–55.

Schlosberg, D., & Collins, L. B. (2014). From environmental justice to climate justice: Climate change and the discourse of environmental justice. *Wiley Interdisciplinary Reviews: Climate Change, 5*(3), 359–374. https://doi.org/10.1002/wcc.275.

Shackley, S., & Wynne, B. (1995). Global climate change: The natural construction of a science-policy domain. *Science and Public Policy, 22*(4), 218–230.

Shi, L., Chu, E., Anguelovski, I., Aylett, A., Debats, J., Goh, K., Schenk, T., Seto, K. C., Dodman, D., Roberts, D., Roberts, J. T., & VanDeveer, S. D. (2016). Roadmap towards justice in urban climate adaptation research. *Nature Climate Change, 6*, 131–137.

Sovacool, B. K. (2018). Bamboo beating bandits: Conflict, inequality, and vulnerability in the political ecology of climate change adaptation in Bangladesh. *World Development, 102*, 183–194.

Sovacool, B. K., & Linnér, B.-O. (2015). *The political economy of climate change adaptation*. Basingstoke: Palgrave Macmillan.

Sovacool, B. K., Linnér, B.-O., & Goosite, M. E. (2015). The political economy of climate change adaptation. *Natural Climate Change, 5*(July), 616–618.

Stilwell, F. (2002). *Political economy: The contest of economic ideas*. Oxford: Oxford University Press.

Storbjörk, S., & Hjerpe, M. (2013). 'Sometimes climate adaptation is politically correct': A case study of planners and politicians negotiating climate adaptation in waterfront spatial planning. *European Planning Studies, 22*(11), 2268–2286.

Sweeney, G., Dobson, R., Despota, K., & Zinnbauer, D. (Eds.). (2011). *Global corruption report: Climate change*. London: Earthscan.

Swyngedouw, E. (2009). The antinomies of the postpolitical city: In search of a democratic politics of environmental production. *International Journal of Urban and Regional Research, 33*(3), 601–620.

Swyngedouw, E. (2010a). Apocalypse forever? Post-political populism and the spectre of climate change. *Theory, Culture and Society, 27*(2–3), 213–232.

Swyngedouw, E. (2010b). Whose environment? The end of nature, climate change and the process of post-politicization. *Ambiente & Sociedade, XIV*(2), 69–87.

Talbot, C. (2016). 'Can't govern', 'won't govern'—The strange confluence of governance and neoliberalism. *Public Money & Management, 36*(4), 243–246.

Tangney, P. (2017). *Climate adaptation policy and evidence: Understanding the tensions between politics and expertise in public policy*. London: Routledge.

Tanner, T., & Allouche, J. (2011). Towards a new political economy of climate change and development. *IDS Bulletin, 42*(3), 1–14.

Taylor, M. (2014). *The political ecology of climate change adaptation: Livelihoods, agrarian change and the conflicts of development*. London: Routledge.

Termeer, C., Biesbroek, R., & van den Brink, M. (2012). Institutions for adaptation to climate change: Comparing national adaptation strategies in Europe. *European Political Science, 11*(2), 41–53.

Terry, G. (2009). No climate justice without gender justice: An overview of the issues. *Gender & Development, 17*(1), 5–18.

Walker, K. J. (1994). *The political economy of environmental policy: An Australian introduction*. Sydney: UNSW Press.

Walker, P. A. (2005). Political ecology: Where is the ecology? *Progress in Human Geography, 29*(1), 73–82.

Watts, M. (2000). Political ecology. In E. Sheppard & T. J. Barnes (Eds.), *A companion to economic geography, Oxford, UK & Malden* (pp. 257–274). MA: Blackwell.

Wilbanks, T. J., & Sathaye, J. (2007). Integrating mitigation and adaptation as responses to climate change: A synthesis. *Mitigation and Adaptation of Strategies for Global Change, 12*(5), 957–962.

CHAPTER 7

# Discussion and Conclusions

**Abstract** In closing the volume, this chapter considers the question of the factors that have shaped and directed the evolution of adaptation politics and the studies examining these changes. Three factors are offered. Firstly, the consequences of the growth and expansion of adaptation interests, themes and activity. Secondly, the effect of external developments in the wider world, particularly of a broader political and socio-economic character. Thirdly, the influence of trends in scholarship, notably changes in the dominant approaches in political science. As final remarks, it is suggested that adaptation politics is itself adapting to changing conditions and will continue to do so. In conclusion, it is suggested that adaptation politics should be considered as an important form of adaptation in itself.

**Keywords** Implications · Political developments · Scale · Scholarship

Critics from several different backgrounds have described the late start made by the social sciences in general in addressing climate change (Rayner and Malone 1998; Goodall 2008; Yearley 2009; Bulkeley 2019). Adaptation is regarded as having been downplayed in the early decades of climate change policy and scholarship (e.g., Pelling 2011; Granberg and Glover 2014). It is unsurprising, therefore, that the politics of adaptation has been largely unexplored until comparatively recent times.



L. Glover and M. Granberg,
*The Politics of Adapting to Climate Change*,
https://doi.org/10.1007/978-3-030-46205-5_7

In reviewing the place of politics in the scholarship and practice of adaptation, several categories and concepts from political studies are used to describe a wide-ranging, diverse and voluminous body of material. Invariably, such a review can only be an overview that attempts to define and articulate the dominant and important themes in adaptation. A case is presented for the need and value of identifying the political dimensions of adaptation. As a starting point, politics not only explanation aspects of adaptation outcomes, but is viewed more widely and deeply so as to embrace the embedded politics in processes, settings and wider context. Major political issues in adaptation are explored and an account offered of the political ideologies and political values associated with differing approaches to adaptation and its scholarship. This review concludes with a discussion of the implications of the volume's findings.

While adaptation at first glance might be perceived as a relatively discrete field of research, policy and practice, determining the politics of these activities involves a more diffuse field of interest, necessitating decisions over the 'boundaries' of the subject. For this review, as may be discerned, politics has been addressed in its obvious forms, or relatively so, but clearly other disciplines in the social sciences overlap with politics considerably. As recorded in the volume, the growth of politics in adaptation studies has been rapid and sustained—but beginning from a point in the early adaptation scholarship where it was essentially absent.

Adaptation policy framework processes have traditionally had no place for political considerations and have been poorly equipped to deal with political circumstances. Power, and its asymmetric allocation in society, is a factor enabling the understanding of the predetermining effects of the mediation of biophysical change by political and social factors (e.g., social vulnerability is a social and ecological construct). For the most part, the adaptation discourse focuses on representation with views that this necessitates exchanges between public, community and citizen interests and the policy process, featuring mutual responsiveness and accountability.

In the maturation of adaptation responses, there are discernible influences that have determined the introduction and growth of political aspects that have occurred and how these are understood and interpreted. Three factors that appear to have been influential in this transformation are offered:

- Consequences of the growth of adaptation activity
- Reflections of the broader political and socio-economic developments in the contemporary era, and
- Changes in the dominant approaches in political science.

## Increasing Scale of Activity

Adaptation to climate change, as identified earlier, and its associated research and scholarship began from a very small base; much of the early activity was derived from other fields, most notably disaster response and more recently, merged with the development discourse as well. Evolution of the adaptation field, if it may be labelled as such, has seen its growth in many different respects, most notably in the extent of policy initiatives and practice, and the corresponding expansion of scholarship. There are many calls for further 'mainstreaming' of adaptation into the development discourse. Climate change impacts research has played a major role in this response, identifying and evaluating a greatly expanded range of natural and social phenomena and processes undergoing change, or forecast to do so.

Concomitant with these changes in scale is the broadening of awareness of the needs for adaptation across society. It follows that the range of potential stakeholders in adaptation has grown and the demands for public sector response has increased. Public sector institutions have been established and/or expanded to satisfy the needs for adaptation. Competition for the public resources made available for adaptation responses is increasing and public sector budgeting will require trade-offs against other spending areas. Accordingly, the consequences of the changes in scale of the adaptation response have increasingly evoked the politics inherent in such a scenario. There are more and varied stakeholders, greater competition for resource allocations, trade-offs being made between competing sectors, locations and interests, differing political values being evoked—all of which contribute to a growing intensification of the political issues and contests involving adaptation. Specialist institutions also play a role by providing 'arenas' in which such political activities can take place.

Therefore, it is concluded that:

- Both the scope and scale of political issues continue to widen, as will its academic analysis, as the scale of adaptation continues to expand and grow in importance
- This expansion in adaptation would necessarily entail a broader engagement of actors (as citizens, governments, businesses and other stakeholders), thereby evoking more varied political interests, values and ideologies, and
- An implication for research into adaptation politics is that the adaptation research field becomes more varied, complicated and contested.

## Political and Socio-Economic Developments

Mitigation of GHG emissions has been a demonstrably important political issue of our time, drawing on and reflecting many of the broader socio-economic and political debates of the contemporary era. In many respects, adaptation has followed this trajectory and also reflects such wider circumstances, due to an increasing:

- Awareness of the associations between mitigation and adaptation, and
- Realisation (and manifestation) of the wider social, economic and environmental impacts of climate change.

Across the developed nations, international financial institutions and beyond, the contemporary political era is that of neoliberalism, featuring market de-regulation, reduced state economic powers, corporatizing and privatizing public entities, use of market-based public policies, labour market deregulation and other features. As explained above, adaptation is a field characterised by 'market failures' and, as such, resolving many adaptation needs is beyond the reach of markets. Opinion may be divided as to the role of neoliberalism in creating barriers and constraints in adaptation, or indeed its place in climate change politics at the broad scale, but market-based approaches to adaptation in public policy are few and far between. An exception may be the role of climate insurance, although this is usually only available to those of relatively higher incomes and may be of limited utility in cases of widespread and extreme impacts. As such, public policy is indispensable for adaptation responses.

This observation should not be interpreted as implying that adaptation scholarship, commentary, plans and practices lie outside of capitalist economic relations, as this is clearly not the case. Rather, it can be argued that it is the realisation of threats to the economic interests of states, regions, corporations and communities/households that has prompted much of the increased adaptation. Providing adaptation, however, is an activity requiring public sector intervention.

Therefore, it is concluded that:

- These political and socio-economic developments may test/challenge existing adaptation institutions (for instance, technical/instrumental approaches to spatial planning may be challenged by approaches based in more explicit political values)
- Adaptation governance will increasingly reflect both established and emerging political controversy and debate, and
- Researchers from relevant disciplines and fields that have previously been unengaged in adaptation studies will be increasingly attracted to adaptation issues, problems and enquiries.

## Politics Scholarship

Another influence is that of the dynamics of political scholarship itself, in which the focus and interests of scholarship shift with the trends within the discipline. Through the era of climate change as a political issue (circa 1990- onwards), the dominant doctrine in academic political science has been that of 'new institutionalism'. Indeed, its rise is largely coincident with neoliberalism in public policy and that of new public management in public administration. New institutionalism, in brief, concerned giving greater attention to the role, place and influence of political institutions (as enduring formal and informal norms, rules and practices), especially their social influences.

Institutions are regarded as 'structures of meaning', shaping, determining and legitimising behaviours. These can also serve as resources that add capacities to social groups, but also function to prescribe and delimit social activity and to impose sanctions. (Institutionalism contrasts with other political explanations, principally those of rational actors and cultural explanations.) Accordingly, institutionalism was taken up in many studies of adaptation, stressing the role, influence and potential of institutions in organising adaptive responses.

Contemporary adaptation scholarship has gone beyond the confines of institutionalism, especially that of environmental justice, political ecology and political economy. As such, political inquiry has 'brought back in' the larger political, cultural, economic and technological influences such as power, class and social movements (albeit that most scholars concede the role played by institutions in influencing political outcomes).

Therefore, it is concluded that:

- Acknowledging and recognising the politics of adaptation in adaptation scholarship creates opportunities for a more just and equitable adaptation policy agenda
- Political values and ideology that frame adaptation policy and practices need to be recognised and included in this analysis, and
- Future adaptation studies will be much more integrated into other research areas and agendas that have a clear focus and interest in social and political aspects.

## Implications

There are calls for greater 'mainstreaming' of adaptation and related discourses, pointing to the commonality of overall interests and a range of tools, methods and approaches offered to assist this process between disaster risk reduction, sustainable development and adaptation. For the most part, such mergers are presented as a win-win outcome. It is, however, unlikely to be a zero-sum game and, at this stage, far less attention has been given to the potential losses to the adaptation discourse that could arise from several sources, including:

- Differing political interests, political values, stakeholders and institutions in each discourse at whatever sphere of government, international agency or NGO is involved
- Policy, programmes and practice differences in such issues as setting policy goals, data and information gathering, choices made in estimating costs and benefits of adaptation, evaluations of adaptation options, resource allocations, community engagement processes and information dissemination, and

- Engaging with other discourses means engaging with another set of political variables; the politics of the adaptation discourse are invariably altered by such linkages.

Politics is embedded in adaptation research, public policy and practice, whether or not this is known, realised or acknowledged. In many regards, the uptake of adaptation has been frequently and largely without regard to its social and environmental implications, something that more recent studies—as reviewed here—have revealed. An under-appreciated virtue of political tools and concepts is that they can provide scholars and analysts with insights into fundamental aspects of adaptation concepts, language, policies, plans and programs that have been hitherto overlooked or ignored in a field dominated by technical, engineering, scientific and economic discourses. Politics provides the means to interrogate adaptation with a view to the implications of its design and intent for social goals, such as justice, the future of the environment, democracy and other conditions.

Does adaptation politics have a distinct identity or can it be understood in its entirety with existing policy theories and ideologies? At this point, adaptation politics can be fitted into existing political explanations, albeit that nearly all scholarship and research efforts have been directed at identifying its characteristics, rather than locating any unique political features of the discourse. This is not to argue that adaptation politics may not or could not develop so as to achieve a singular political identity. If this does occur, it would most likely probably happen in conjunction, and integrated with, two existing and related discourses—those of disaster studies and development studies. In any event, adaptation politics will continue to evolve in response to changing external conditions and from its own dynamic features; adaptation politics itself will continue to adapt to change.

Adaptation entails basic issues and concepts central to our notions of politics, such as power, knowledge, authority and legitimacy: Why do we undertake adaptation? What is to be protected? What is to be left unprotected? Who benefits and who loses from adaptation? Who carries the costs of adaptation? Will some regions or places benefit at the expense of others? How are natural values valued? Who 'speaks' for Nature? How is the future, and how are future generations, valued? Who makes decisions? What institutions are involved? What counts as legitimate knowledge? What are adaptation priorities and how are these set?

Answering questions such as these draws on political concepts, values and ideologies. Politics, as a discipline, has uncovered social and environmental values embedded in adaptation and that, in part, is its role. Not only has politics provided the means to address social inquiries of adaptation, but it has also prompted new inquiries.

Politics will ultimately determine society's adaptation response and the more widely this is appreciated, the better society and environmental interests can be served by adaptation measures. Scholarship, therefore, into the politics of adaptation is ongoing, important and necessary in order to reveal and build knowledge, understand its implications and inform policy and society at large. Adaptation policies, programs and activities will be improved through the realisation that adaptation politics are an important component of adaptation; these politics constitute an adaptive response in their own right.

## References

Bulkeley, H. (2019). Navigating climate's human geographies: Exploring the whereabouts of climate politics. *Dialogues in Human Geography, 9*(1), 3–17.

Goodall, A. H. (2008). Why have the leading journals in management (and other social sciences) failed to respond to climate change? *Journal of Management Inquiry, 17*(4), 408–420.

Granberg, M., & Glover, L. (2014). Adaptation and maladaptation in Australian national climate change policy. *Journal of Environmental Policy & Planning, 16*(2), 147–159.

Pelling, M. (2011). *Adaptation to climate change: From resilience to transformation*. London: Routledge.

Rayner, S., & Malone, E. L. (Eds.). (1998). *Human choice and climate change: Vol. I: The Societal Framework*. Columbus: Batelle Press.

Yearley, S. (2009). Sociology and climate change after Kyoto: What roles for social science in understanding climate change? *Current Sociology, 57*(3), 389–405.

# Index





L. Glover and M. Granberg,
*The Politics of Adapting to Climate Change*,
https://doi.org/10.1007/978-3-030-46205-5

## O

## P

## S

T

The manufacturer's authorised representative in the EU is Springer
Nature Customer Service Centre GmbH, Europaplatz 3, 69115 Heidelberg,
Germany. If you have any concerns regarding our products, please
contact ProductSafety@springernature.com

Printed and bound by CPI Group (UK) Ltd, Croydon, CR0 4YY
30/09/2026
02225649-0001